Collins

Edexcel

International GCSE

Chemistry

Chris Sunley

Teacher Pack

William Collins' dream of knowledge for all began with the publication of his first book in 1819. A self-educated mill worker, he not only enriched millions of lives, but also founded a flourishing publishing house. Today, staying true to this spirit, Collins books are packed with inspiration, innovation and practical expertise. They place you at the centre of a world of possibility and give you exactly what you need to explore it.

Collins. Freedom to teach

Published by Collins
An imprint of HarperCollins*Publishers*
77 – 85 Fulham Palace Road
Hammersmith
London
W6 8JB

MIX
Paper from
responsible sources
FSC
www.fsc.org
FSC™ C007454

FSC™ is a non-profit international organisation established to promote the responsible management of the world's forests. Products carrying the FSC label are independently certified to assure consumers that they come from forests that are managed to meet the social, economic and ecological needs of present and future generations, and other controlled sources.

Find out more about HarperCollins and the environment at
www.harpercollins.co.uk/green

10 9 8 7 6 5 4 3 2 1

ISBN-13 978 0 00 745004 6

British Library Cataloguing in Publication Data
A Catalogue record for this publication is available from the British Library

Commissioned by Letitia Luff and Rebecca Richardson
Project managed by Jim Newall and Caroline Green
Formatted by Jim Newall and Lauren Bourque
Manuscript review and answer check by Rob Wensley
Edited by Jim Newall and Lauren Bourque
Proofread by Lauren Bourque and Grace Glendinning
Word template design by Jerry Fowler
New illustrations by Gray Publishing and Kathy Baxendale
Cover design by Angela English
Production by Rebecca Evans
Printed and bound by Martins the Printers

Health and Safety: All practicals in this book have been safety checked (but not trialled) by CLEAPSS (www.cleapss.org.uk) and the publisher and authors have made every reasonable effort to ensure that the experiments and activities in this book are safe when conducted as instructed. However, the publishers assume no responsibility for any damage or injury caused or sustained while performing the experiments or activities in this book to the full extent permitted by law.

CLEAPSS

Parents, guardians, and/or teachers should supervise students who undertake the experiments and activities in this book, and carry out full risk assessments before carrying out any practical work.

CLEAPSS is an advisory service providing support in science and technology for a consortium of local authorities and their schools including establishments for pupils with special needs.

Acknowledgements
Every effort has been made to trace copyright holders and to obtain their permission for the use of copyright materials. The publishers will gladly receive any information enabling them to rectify any error or omission at the first opportunity.

Contents

Introduction **5**

Suggested teaching sequence **7**

Matching chart to double and single awards **9**

Section 1: Principles of Chemistry **13**

a) States of matter **15**
- C1a.1 Orientation 16
- C1a.2 Kinetic theory of matter 17
- C1a.3 Changes of state 19
- C1a.4 Consolidation and summary 20

b) Atoms **21**
- C1b.1 Orientation 22
- C1b.2 Diffusion 23
- C1b.3 Elements, atoms, molecules, compounds and mixtures 25
- C1b.4 Separation of mixtures 27
- C1b.5 Consolidation and summary 31

c) Atomic structure **33**
- C1c.1 Orientation 34
- C1c.2 Sub-atomic particles 36
- C1c.3 Relative atomic mass and the mole 37
- C1c.4 Electron arrangements 39
- C1c.5 Consolidation and summary 43

d) Relative formula masses and molar volumes of gases **45**
- C1d.1 Orientation 46
- C1d.2 Relative formula masses 47
- C1d.3 Molar volumes of gases 49
- C1d.4 Consolidation and summary 50

e) Chemical formulae and chemical equations **54**
- C1e.1 Orientation 56
- C1e.2 Finding the formula of magnesium oxide 57
- C1e.3 Empirical and molecular formulae 60
- C1e.4 Finding the formula of copper(II) oxide 61
- C1e.5 Writing chemical equations 65
- C1e.6 Reacting masses 67
- C1e.7 Consolidation and summary 72

f) Ionic compounds **75**
- C1f.1 Orientation 76
- C1f.2 The formation of ions 77
- C1f.3 Properties of ionic compounds 80
- C1f.4 Consolidation and summary 81

g) Covalent substances **83**
- C1g.1 Orientation 84
- C1g.2 The formation of covalent molecules 85
- C1g.3 Properties of covalent compounds 88
- C1g.4 Diamond and graphite 89
- C1g.5 Consolidation and summary 90

h) Metallic crystals **92**
- C1h.1 Orientation 93
- C1h.2 Metallic bonding 94
- C1h.3 Consolidation and summary 95

i) Electrolysis **97**
- C1i.1 Orientation 98
- C1i.2 Electrolysis of molten electrolytes 100
- C1i.3 Electrolysis of aqueous electrolytes 101
- C1i.4 Quantitative electrolysis 107
- C1i.5 Consolidation and summary 110

Exam-style questions answers 112

Section 2: Chemistry of the Elements **113**

a) The Periodic Table **115**
- C2a.1 Orientation 116
- C2a.2 The arrangement of the elements 117
- C2a.3 Characteristic properties of metals and non-metals 118
- C2a.4 Consolidation and summary 124

b) Group 1 elements **126**
- C2b.1 Orientation 127
- C2b.2 Reactions of the Group 1 elements 128
- C2b.3 Consolidation and summary 130

c) Group 7 elements **132**
- C2c.1 Orientation 133
- C2c.2 Displacement reactions of the halogens 134
- C2c.3 Hydrogen chloride 138
- C2c.4 Consolidation and summary 140

d) Oxygen and oxides **142**
- C2d.1 Orientation 144
- C2d.2 Determining the percentage composition of oxygen in the air 145
- C2d.3 The preparation and properties of oxygen 147
- C2d.4 The preparation and properties of carbon dioxide 151
- C2d.5 Carbon dioxide as a greenhouse gas 154
- C2d.6 Consolidation and summary 156

e) Hydrogen and water **158**
- C2e.1 Orientation 159
- C2e.2 The preparation of hydrogen 160
- C2e.3 Water 164
- C2e.4 Consolidation and summary 166

f) Reactivity series **168**
- C2f.1 Orientation 170
- C2f.2 Displacement reactions 171
- C2f.3 Limiting corrosion 176
- C2f.4 Consolidation and summary 180

g) Tests for ions and gases **181**
C2g.1 Orientation 183
C2g.2 Identifying cations 184
C2g.3 Identifying anions 188
C2g.4 Identifying gases 192
C2g.5 Consolidation and summary 196
Exam-style questions answers 198

Section 3: Organic Chemistry **199**
a) Alkanes **201**
C3a.1 Orientation 202
C3a.2 The structure of alkanes 204
C3a.3 The chemical properties of alkanes 206
C3a.4 Consolidation and summary 207
b) Alkenes **209**
C3b.1 Orientation 210
C3b.2 The chemical properties of alkenes 211
C3b.3 Consolidation and summary 213
c) Ethanol **215**
C3c.1 Orientation 217
C3c.2 The manufacture of ethanol 218
C3c.3 Reactions of ethanol 220
C3c.4 Consolidation and summary 224
Exam-style questions answers 226

Section 4: Physical Chemistry **227**
a) Acids, alkalis and salts **229**
C4a.1 Orientation 231
C4a.2 Acids, bases, alkalis and salts 235
C4a.3 Making soluble salts 236
C4a.4 Making insoluble salts 243
C4a.5 Consolidation and summary 245
b) Energetics **248**
C4b.1 Orientation 250
C4b.2 Measuring energy transfers (1) 254
C4b.3 Measuring energy transfers (2) 261
C4b.4 Energy level diagrams and bond breaking and bond forming 265
C4b.5 Consolidation and summary 267
c) Rates of reaction **269**
C4c.1 Orientation 271
C4c.2 Monitoring the rate of a reaction 274
C4c.3 Surface area of solids 277
C4c.4 Concentration of solutions 280
C4c.5 Temperature 284
C4c.6 Use of a catalyst 287
C4c.7 Consolidation and summary 290
d) Equilibria **292**
C4d.1 Orientation 293
C4d.2 Changing the position of equilibrium 295
C4d.3 Consolidation and summary 298
Exam-style questions answers 300

Section 5: Chemistry in Society **301**
a) Extraction and uses of metals **303**
C5a.1 Orientation 304
C5a.2 Aluminium: extraction and uses 308
C5a.3 Iron: extraction and uses 310
C5a.4 Consolidation and summary 313
b) Crude oil **315**
C5b.1 Orientation 317
C5b.2 The fractional distillation of crude oil 318
C5b.3 Cracking crude oil fractions 322
C5b.4 The combustion of fuels and acid rain 326
C5b.5 Consolidation and summary 327
c) Synthetic polymers **329**
C5c.1 Orientation 330
C5c.2 Addition polymerisation 331
C5c.3 Condensation polymerisation 335
C5c.4 Consolidation and summary 337
d) Industrial manufacture of chemicals **339**
C5d.1 Orientation 341
C5d.2 The manufacture of ammonia 342
C5d.3 The manufacture of sulfuric acid 343
C5d.4 The manufacture of sodium hydroxide and chlorine 344
C5d.5 Consolidation and summary 346
Exam-style questions answers 348

Scheme of work **349**

Introduction

Welcome to the Collins Edexcel International GCSE Chemistry Teacher Pack, which has been written by experienced authors and teachers and fully safety checked by CLEAPSS to help you deliver an effective and successful International GCSE Chemistry course.

This pack comprises two components – this printed book and a CD–ROM.

Overview

Printed Book

The **printed book** mirrors the Student Book topics and includes:

- **Activity plans** with resource lists, learning outcomes and objectives, and detailed guidance.
- **Practical activities** including teacher demonstrations and student experiments.
- **Worksheets** for students with instructions to follow and questions to answer.
- **Technician's notes** so practicals can be safely planned and executed.
- **Answers** to all the questions in the Student Book.
- **Scheme of Work** that maps all practical activities and shows at a glance all the topics linked with their learning outcomes and Edexcel specification references.
- Matching chart to the **Double Award** specification.

CD–ROM

The **CD–ROM** offers everything in the printed book in MS Word format so that you can tailor lessons and activities. The content has been arranged in folders by Section and then by Topic, with all the materials you need so that you can quickly and easily prepare for your lessons.

It also includes:

- **Editable versions** of the Activity plans, Worksheets, and Scheme of Work in the printed book.
- **Technician's notes** with lists of equipment, resources and set-up information are included with the relevant lesson topic. All the notes for each section are also collated in one document you can hand to your technician.
- Printable **Checklists** students can use to map their own progress.
- One document containing all the **Student Book answers**.

Activity plans

The Teacher Pack has been matched topic by topic to the Student Book to allow easy cross referencing between the books. Each topic is introduced with a brief introduction, links to other topics and an overview of the activities in an easy-to-navigate table. Activity plans (called 'teacher notes' on the CD-ROM) give guidance on delivering learning episodes that can be combined for longer lessons to give you complete flexibility and control.

Activity plans cover the following:

- **Learning objectives and outcomes** linked to the specification and the student book.
- **Common misconceptions** that students hold which may need addressing.
- **Resources** you will need for activities.

- A detailed pick-up-and-teach **Approach** for differentiated, varied tasks.
- **Demonstration and practical activity** technician's notes so you can easily check you have everything you need.
- **Answers** to all the questions in the Student Book.

Activity plans are supported by **Worksheets, Technician's notes** and all **Answers**.

Worksheets

Worksheets provide engaging activities for students. Those that deal with practical work give step-by-step guidance for students to ensure that they carry out their investigations safely and successfully.

All Worksheets are available in editable MS Word format on the CD–ROM as well as in one printable document containing all the Worksheets in each section.

Technician's notes

Detailed **Technician's notes** are provided for every lesson with lists of resources needed for practical activities and/or demonstrations, set-up instructions and safety notes.

For Today's Safety Information, which will need to be checked before setting up any practical activity or demonstration, please refer to your employers guidelines or visit: www.cleapss.org.uk

Technician's notes appear within the lesson plans for your reference, and as separate printable files on the CD–ROM.

Answers

Every activity plan in this book includes the **answers** to the associated questions in the Student Book. Answers to the **Exam-style questions** are given at the end of every section.

In addition, all the answers are collated in a separate document on the CD–ROM for easy printing.

Checklists

Editable checklists that students can use to record their progress are provided on the CD–ROM. These have the same content as those at the end of each topic in the Student Book.

Scheme of Work

The detailed **Scheme of Work** at the back of this book provides a comprehensive overview of the Edexcel International GCSE course, matching each lesson to the learning outcomes in the specification. It also shows at a glance where the practical activities and demonstrations occur so that you can plan ahead.

An **editable version of this Scheme of Work** can also be found on the CD–ROM.

Suggested teaching sequence

Introduction

The specification content is divided into five sections, each containing a number of different topics:

1. Principles of chemistry
2. Chemistry of the elements
3. Organic chemistry
4. Physical chemistry
5. Chemistry in Society

When you are determining the teaching sequence for the course, you should keep in mind these important considerations:

- There should be a logical progression in the development of concepts and knowledge in order to aid students' understanding.
- The course should have a 'storyline' so that wherever possible the sequence can be easily explained to students.
- Consider carefully how to introduce the more difficult concepts. For example, ensure that they are not introduced too early in the course, which may de-motivate students, or put in a sequence where they all come one after another.
- Maintain the emphasis on practical and investigative work throughout the course.
- You will need to allocate different amounts of teaching time to different topics, depending on levels of difficulty or extent of the content.
- It is important to retain some flexibility so that you can revisit difficult ideas or apply new ideas in different contexts.

Suggested teaching sequence

The sequence is mapped over 5 terms on the assumption that you will spend much of the 6th term on revision. Where the course is started early or where the time allocated each week is either generous or limited, you will need to make adjustments.

First year, 1st term

1a States of matter

1b Atoms

1c Atomic structure

1f Ionic compounds

1g Covalent substances

4a Acids, alkalis and salts

1e Chemical formulae and equations (part 1)

2d Oxygen and oxide

2e Hydrogen and water

First year, 2nd term

2a The Periodic Table

2b Group 1 elements

2c Group 7 elements

4c Rates of reaction

1d Relative formula masses and molar volumes of gases

1e Chemical formulae and chemical equations (part 2)

First year, 3rd term

1h Metallic crystals

2f Reactivity series

1i Electrolysis

2g Tests for ions and gases

Second year, 1st term

4b Energetics

4d Equilibria

3a Alkanes

3b Alkenes

3c Ethanol

Second year, 2nd term

5a Extraction and uses of metals

5b Crude oil

5c Synthetic polymers

5d The industrial manufacture of chemicals

Notes

1. Topic 1e *Chemical formulae and chemical equations* has been divided. This is to spread out the introduction of a substantial amount of quantitative chemistry.

2. The Chemistry in Society topics have been all grouped together because they provide a way of consolidating and extending previous learning.

Matching chart to Double and Single Awards

This chart matches the statements in the Edexcel Double and Single Awards to the content of the Student Book. A shaded box in a column indicates that the topic is not needed in the Double Award.

The numbers in the columns correspond to the statement numbers in the specification.

Section 1 Principles of chemistry

Topic	Single Award	Double Award	Pages
a) States of matter	1.1	1.1	10–12
	1.2	1.2	12, 13
	1.3	1.3	14
b) Atoms	1.4	1.4	18–20
	1.5	1.5	21, 22
	1.6	1.6	21–23
	1.7	1.7	23–25
	1.8	1.8	25
c) Atomic structure	1.9	1.9	28, 29, 34
	1.10	1.10	29
	1.11	1.11	30–32
	1.12	1.12	33
	1.13	1.13	30, 108, 109
	1.14	1.14	34, 35
	1.15	1.15	36, 37, 112
d) RFM and molar volumes of gases	1.16	1.16	41, 42
	1.17	1.17	32, 33, 42, 43
	1.18		32, 33, 42, 43
	1.19	1.18	32, 33, 45
	1.20		44, 45
e) Chemical formulae and equations	1.21	1.19	56–60
	1.22	1.20	56, 206
	1.23	1.21	50–54
	1.24	1.22	53–55
	1.25	1.23	61, 62
	1.26		63, 64
	1.27	1.24	62, 63, 214
f) Ionic compounds	1.28	1.25	68
	1.29	1.26	70, 96
	1.30	1.27	71
	1.31	1.28	112
	1.32	1.29	69, 70
	1.33	1.30	71, 72
	1.34	1.31	71, 72
	1.35		72, 73
	1.36		71, 73
	1.37		71
g) Covalent substances	1.38	1.32	76
	1.39	1.33	77
	1.40	1.34	77–79
	1.41	1.35	80
	1.42	1.36	80
	1.43	1.37	81

	1.44		81
	1.45		81
h) Metallic crystals	1.46	1.38	85
	1.47	1.39	85, 86
i) Electrolysis	1.48	1.40	91
	1.49	1.41	100
	1.50	1.42	91
	1.51	1.43	90, 91
	1.52	1.44	92
	1.53	1.45	92
	1.54		93–97
	1.55	1.46	92, 94
	1.56		98
	1.57		98

Section 2 Chemistry of the elements

Topic	Single Award	Double Award	Pages
a) The Periodic Table	2.1	2.1	109
	2.2	2.2	109, 110
	2.3	2.3	111, 137
	2.4	2.4	112
	2.5	2.5	112
b) Group 1 elements	2.6	2.6	117, 118
	2.7	2.7	116
	2.8		112, 116
c) Group 7 elements	2.9	2.8	125
	2.10	2.9	124, 130
	2.11	2.10	129
	2.12	2.11	129
	2.13	2.12	124
	2.14	2.13	126, 128
	2.15	2.14	126
d) Oxygen and oxides	2.16	2.15	134
	2.17	2.16	135, 138
	2.18	2.17	136
	2.19	2.18	136, 137
	2.20	2.19	138, 139
	2.21	2.20	139
	2.22	2.21	139
	2.23	2.22	140
	2.24	2.23	140
e) Hydrogen and water	2.25	2.24	145, 146
	2.26	2.25	146
	2.27	2.26	147
	2.28	2.27	147
f) Reactivity series	2.29	2.28	151
	2.30	2.29	151, 152
	2.31	2.30	152–154
	2.32	2.31	152
	2.33	2.32	152
	2.34	2.33	155
	2.35	2.34	155

Topic	Single Award	Double Award	Pages
	2.36	2.35	155
g) Tests for ions and gases	2.37	2.36	160–162
	2.38	2.37	162–164
	2.39	2.38	164, 165

Section 3 Organic chemistry

Topic	Single Award	Double Award	Pages
a) Introduction	3.1	3.1	176, 177, 179, 186
b) Alkanes	3.2	3.2	177, 178
	3.3	3.3	177, 179, 180
	3.4	3.4	178, 181
	3.5	3.5	181
c) Alkenes	3.6	3.6	186
	3.7	3.7	185, 187
	3.8	3.8	186
d) Ethanol	3.9		196
	3.10		193, 194, 195
	3.11		196
	3.12		196

Section 4 Physical chemistry

Topic	Single Award	Double Award	Pages
a) Acids, alkalis and salts	4.1	4.1	206, 207
	4.2	4.2	207
	4.3	4.3	207
	4.4	4.4	208, 209
	4.5	4.5	209
	4.6	4.6	210
	4.7	4.7	211–213
	4.8	4.8	214, 215
	4.9	4.9	211, 213, 216, 217
b) Energetics	4.10	4.10	222
	4.11	4.11	223, 224, 227
	4.12		225
	4.13	4.12	225
	4.14	4.13	226, 228
	4.15	4.14	228, 229
	4.16		229, 230
c) Rates of reaction	4.17	4.15	236–240, 242, 243
	4.18	4.16	237, 240–248
	4.19	4.17	226, 237, 238
	4.20	4.18	241–247
	4.21	4.19	246, 247
d) Equilibria	4.22	4.20	252, 253
	4.23	4.21	253
	4.24	4.22	253, 254
	4.25	4.23	254–257

Section 5 Chemistry in society

Topic	Single Award	Double Award	Pages
a) Extraction and uses of metals	5.1	5.1	156, 268, 269
	5.2	5.2	269, 270
	5.3	5.3	269
	5.4	5.4	272, 273, 273
	5.5	5.5	270, 274, 275
b) Crude oil	5.6	5.6	282
	5.7	5.7	281, 282
	5.8	5.8	281
	5.9	5.9	282, 283
	5.10	5.10	285, 286
	5.11	5.11	286, 287
	5.12	5.12	287, 288, 289
	5.13	5.13	283, 284
	5.14	5.14	284, 285
c) Synthetic polymers	5.15	5.15	295
	5.16	5.16	296
	5.17	5.17	??
	5.18	5.18	296
	5.19	5.19	298
	5.20		297
	5.21		297
d) The industrial manufacture of chemicals	5.22	5.20	302, 303
	5.23	5.21	303
	5.24	5.22	303
	5.25	5.23	303, 304
	5.26		304, 305
	5.27		305, 306
	5.28		305
	5.29		307, 308
	5.30		307
	5.31		308

Section 1. Principles of chemistry

Contents

a) States of matter

b) Atoms

c) Atomic structure

d) Relative formula masses and molar volumes of gases

e) Chemical formulae and chemical equations

f) Ionic compounds

g) Covalent substances

h) Metallic crystals

i) Electrolysis

j) Exam-style questions

Overview of the section

This section provides the foundations for the rest of the course. The early topics deal with the states of matter and the structure of the atom. Chemical equations are then introduced, focusing on how equations provide quantitative information about reactions. This is likely to be students' first experience of quantitative chemistry.

Later on, the section deals with the structure of metals and the structure, bonding and characteristic properties of ionic and covalent substances, including the electrolysis of ionic compounds.

You will need to teach many of these topics early in the International GCSE course (see the suggested teaching sequence). However, to maintain the practical and investigative elements, it is important not to try and teach this entire section before moving on to topics in the other four sections.

Most students starting the International GCSE course will have had some previous experience of chemistry. What may pose problems is that their experiences may be quite different, and their recall and level of knowledge may vary considerably. Consequently, it is important that at the beginning of each topic you spend time finding out exactly what students remember from their previous study of chemistry.

Because many of the topics in this section will be taught at the beginning of the course, it is important to use the 'orientation' stage to establish the extent of students' prior knowledge.

Starting points

The Student Book provides a 'section opener', a double-page spread which sets the scene for students (see pages 8–9). It is structured in this way:

- an overview providing details of the areas of study
- six questions, the first three relating to aspects of chemistry that should be reasonably familiar to most students, and the final three relating directly to some of the topics to be studied
- a list of section contents.

The questions provide a structure for introducing the section.

1. What is an atom?
2. Do you know the names of any of the particles that are found in an atom?
3. What is the name of the particle formed when two atoms combine together?
4. You will be learning about the states of matter. Do you know what these states are?

5. One type of chemical bonding you will study is called ionic bonding. Do you know, or can you find out, what an ion is?

6. Diamond and graphite are both covalent substances. They contain the same atoms but have very different structures and properties. What do you know about the properties of diamond and graphite?

The section opener has two main purposes:

- to acknowledge the student's prior learning and to value it
- to provide a benchmark against which future learning can be compared.

The six questions or 'starting points' in the Student Book can be used in a number of different ways to introduce the section:

- You could ask students to consider the questions as an introductory homework task.
- You could put students into groups to share their own ideas and understanding and then report back to the whole class.
- Students could be given access to the Internet, preferably with a tight timescale, to find out the information required.

You could then use a spider chart or other form of wall chart to summarise everybody's ideas.

Recording these initial ideas in this way allows you to retain them for reference as the individual topics are developed. In this way, your students' progress in learning can be readily acknowledged.

C1a States of matter

Introduction

This topic may go back over ideas that students have already met, so it does not need to take up much teaching time. Nevertheless, it is important for your students to understand the kinetic theory of matter, which forms the basis of discussions of diffusion in Topic 1b *Atoms* and collision theory in Topic 4c *Rates of reaction*. A short demonstration is included in the first activity.

Links to other topics

Sections	Essential background knowledge	Useful links
1 Principles of chemistry		1b Atoms
4 Physical chemistry		4c Rates of reaction
5 Chemistry in society		5b Crude oil

Topic overview

C1a.1	**Orientation** The purpose of this activity is to recap students' current knowledge and understanding of the three states of matter and the names of the processes involved when substances change states.
C1a.2	**The kinetic theory of matter** This activity introduces the particulate models of solids, liquids and gases. The focus is on the different arrangements and movements of the particles in the three states.
C1a.3	**Changes of state** This activity reinforces the names of the processes which bring about a change in state and relates these changes to the energy involved.
C1a.4	**Consolidation and summary** This activity allows for a quick recap of the key ideas of the topic.

Activity C1a.1 Orientation

Learning objectives

- To classify substances as solid, liquid or gas
- To be familiar with the properties of typical solids, liquids and gases

Learning outcomes

- Know the three states of matter and their shorthand notation (s, l, g)
- Know the typical properties of the three states of matter in terms of volume and shape

Resources

Student Book pages 10–11

Resources for demonstration (see below)

Files on CD-ROM: C1a.1a_tech_notes

Approach

Students should be familiar with classifying solids, liquids and gases. They may also know the names of the processes that cause changes of state. So this activity can be relatively short.

Ask students to list four solids, four liquids and four gases they are familiar with. Emphasise that substances with the same classification can have quite different properties. For example: a candle and a rod of metal (hardness); water and liquid honey (viscosity); air and bromine gas (colour).

Use Student Book page 11 to summarise the different properties of solids, liquids and gases. Explain that these are generally true, but that students need to be careful: for example, sand is obviously a solid, but it takes the shape of any container it is put in.

Finally, ask students to name some of the processes involved in changing states, such as melting, boiling (evaporation), freezing, etc.

Technician's notes

You will need the following resources for a demonstration.

Be sure to check the latest safety notes on these resources before proceeding.

a candle
a metal bar (similar diameter to metal bar)
a jar of liquid honey
a gas jar of bromine gas (in fume cupboard)

Activity C1a.2 Kinetic theory of matter

Learning objective

- To introduce the kinetic theory of matter

Learning outcomes

- Know the arrangement of particles in solids, liquids and gases
- Understand the movement and energy of the particles in solids, liquids and gases

Common misconceptions

The particles in a liquid are not touching, that is, the arrangement of particles in a liquid is closer to that of a gas than to that of a solid.

Resources

Student Book pages 11–12

Worksheet C1a.2a The particulate nature of matter

Files on CD-ROM: C1a.2a_worksheet

Approach

Use Student Book page 12 to illustrate the particle arrangements in solids, liquids and gases. Emphasise the arrangements and movements of the particles in each model.

Ask students to speculate on the relative strength of the forces between the particles in the three models and use their ideas to explain why gases expand on heating more than solids and liquids.

Finally, use Worksheet C1a.2a to check that students understand the key differences between the three states. (Note: this worksheet can also be used in Topic 4c *Rates of reaction* as part of the orientation activity.)

Answers

Page 11

1. (l).

2. Only the solid state has a fixed shape.

3. Fine sand will pour or flow like a liquid; it takes the shape of the container it is poured into (although under a microscope you would see gaps at the surface of the container).

C1a.2a The particulate nature of matter

Cut out the cards and then sort them into piles of 'True', 'False' or 'Not sure'.

1 In a solid the particles are not moving.	2 The particles in a gas are very far apart.	3 In a liquid, the particles will be constantly bumping into each other.
4 The particles in a gas are moving at high speed.	5 The particles in a solid are held together by various types of forces.	6 Solid A has a higher melting point than solid B so the forces holding the particles together in A are stronger than those in B.
7 As a liquid is heated, the particles move faster.	8 As a solid is heated, the particles stay still.	9 The forces between gas particles are very small.
10 When a liquid evaporates particles escape from the surface of the liquid.	11 When a solid melts and forms a liquid, the particles usually move further apart.	12 When gas particles mix together, this is known as dispersion.

Activity C1a.3 Changes of state

Learning objectives

- To reinforce the names of the main processes involved in changes of state
- To use the kinetic theory of matter to explain the changes that occur

Learning outcomes

- Be able to describe how the three states of matter can be interconverted and know the names for these processes
- Understand how the interconversions of melting, boiling, evaporation, freezing and condensation take place by describing changes in the arrangement, movement and energy of the particles involved

Common misconceptions

Some students think that condensation is not a process but a 'mist of water' found on cold surfaces.

Resources

Student Book pages 12–14

Approach

Use Student Book page 13 to illustrate melting, boiling, freezing and condensing. Emphasise the similarities and differences between evaporation and boiling.

Use the sequence ice–water–steam on page 14 to reinforce the terminology. Use kinetic theory to explain the changes that occur. Do the same for the reverse sequence from steam back to ice.

Ask students to answer the questions on page 13.

Answers

Page 13

1. The particles in a solid vibrate.
2. Water particles are held together more strongly in liquid water.
3. Evaporation is the process in which faster-moving particles escape from the surface of a liquid.
4. A solid changes to a liquid at its melting point.

Activity C1a.4 Consolidation and summary

Learning objectives

- To review the key learning points of the topic
- To test understanding through answering questions

Learning outcomes

- Be familiar with the knowledge and understanding summarised in the End of Topic Checklist.
- Be able to apply this knowledge and understanding by answering the End of Topic Questions.

Resources

Student Book pages 16–17

Approach

This exercise can be done individually or in small groups. Give priority to listening to students' ideas and explanations. Finally, use the feature on the states of matter (Student Book page 15) to discuss plasma as a potential fourth state of matter.

Answers

End of Topic Questions mark scheme

Question	Correct answer	Marks
1	Gas	1 mark
2	The particles in a liquid are close together, often touching. The particles can move around and are not in a fixed position.	1 mark 1 mark
3	Solid	1 mark
4	The forces between the particles (atoms) in aluminium are greater than those between the particles (atoms) in sodium.	1 mark
5 a)	Melting	1 mark
5 b)	Evaporation or boiling	1 mark
5 c)	Condensation	1 mark
5 d)	Freezing or solidification	1 mark
6	The boiling point	1 mark
7	The particles in a liquid are constantly moving even at a low temperatures in the polar regions. Water particles can break away from the surface (evaporate) and form water vapour.	1 mark 1 mark
8	The student's statement is correct. In both solids and liquids the particles are quite close together – in a gas the particles are much further apart. The speed of movement of the particles in solids and liquids are much more similar – in a gas the particles are moving at very high speeds.	1 mark 1 mark
	Total:	14 marks

C1b Atoms

Introduction

You may find that this topic overlaps quite a bit with what students have already learned. After the orientation activity you will find it easier to judge how much teaching time to devote to the topic. This topic focuses first on diffusion and then on understanding the nature of atoms, molecules, elements, compounds and mixtures. All of this underpins much of the work encountered in the rest of the course.

Links to other topics

Sections	Essential background knowledge	Useful links
1 Principles of chemistry	1a States of matter	1e Chemical formulae and chemical equations 1f Ionic compounds 1g Covalent substances
2 Chemistry of the elements		2a The Periodic Table

Topic overview

C1b.1	**Orientation** Students will already have some knowledge of atoms, molecules, compounds and ways of separating mixtures. The purpose of this activity is to assess their prior knowledge, identify any misconceptions and then decide how much teaching time you will need to spend on the topic.
C1b.2	**Diffusion** This activity involves a series of simple diffusion experiments, some of which can be demonstrated and some undertaken by the students themselves. Evidence of the constant movement of gas particles can be provided through the observation of Brownian motion in a smoke cell.
C1b.3	**Elements, atoms, molecules, compounds and mixtures** This activity makes the link between a number of different chemical terms which, at times, are interchangeable and so can be confusing. It is a short activity intended to clarify the definitions of the terms and to provide students with the opportunity to apply their knowledge.
C1b.4	**Separation of mixtures** This activity is intended to make sure that the students are familiar with a number of separation processes: distillation, fractional distillation, filtration, crystallisation and chromatography. These aspects may be very familiar to students and so the activity may be quite short.
C1b.5	**Consolidation and summary** This activity provides a quick recap of the key ideas of the topic.

Activity C1b.1 Orientation

Learning objectives

- To review knowledge of atoms, molecules and compounds
- To review knowledge of the separation of mixtures

Learning outcomes

- Know that an atom is the smallest part of an element that can exist on its own
- Know that mixtures can be separated in a number of different ways

Resources

Student Book pages 18, 23–25

Files on CD–ROM: C1b.1a_tech_notes

Resources for demonstration (see below)

Approach

Divide students into small groups. Ask them to write down definitions of the terms element, atom, molecule and compound. Discuss their responses and, if appropriate, display the definitions on the board so they can be reviewed later in the topic.

At this stage you do not need to identify which are the most accurate definitions. This will be covered in Activity C1b.3.

Set up demonstrations of distillation, fractional distillation, filtration, crystallisation and chromatography. As students observe each process, ask them to identify which states of matter are being separated in each process. You will need to judge whether students have enough prior knowledge to omit Activity C1b.4 altogether.

Technician's notes

You will need the following resources for a demonstration.

Be sure to check the latest safety notes on these resources before proceeding.

simple distillation apparatus (Student Book page 23), with dilute blue washable ink, for example
fractional distillation (Student Book page 23), with ethanol and water, for example
filtration with sand and water, for example
crystallisation (Student Book page 24) with salt solution, for example
chromatography (Student Book page 25) with black soluble ink, for example

Activity C1b.2 Diffusion

Learning objectives

- To observe and discuss a range of diffusion experiments
- To observe and discuss Brownian motion

Learning outcomes

- Be able to describe simple diffusion experiments including dissolving a solid in a liquid and the mixing of gases
- Be able to describe Brownian motion and explain the causes of the motion for pollen grains on water and smoke particles in air

Common misconceptions

Gas particles move relatively slowly because the diffusion of gases is relatively slow.

Resources

Student Book pages 18–20

Files on CD–ROM: C1b.2a_tech_notes, C1b.2a_worksheet

Resources for demonstration and class practical (see below)

Approach

Arrange a mixture of student practical activities and demonstrations. For example:

Student practical activities (these instructions are provided as a separate printable sheet - C1b.2a_worksheet for students on the CD-ROM)

1. Add a crystal of potassium manganate(VII) to still water in a Petri dish. Leave the Petri dish unmoved for at least 20 minutes. Compare observations with those on Student Book page 19.

2. Mix carbon dioxide and air in test tubes and test with lime water (a safer alternative to the hydrogen and air equivalent – see Student Book page 19 Fig. 1.12). You may have to introduce the test for carbon dioxide if students have not seen it before.

A simple carbon dioxide generator (marble chips and dilute hydrochloric acid) can be used to provide a source of the gas. One test tube is filled with carbon dioxide and then connected vertically to a test tube containing air. The tubes are held together for a few minutes and then each tube is tested with lime water to identify carbon dioxide.

The experiment can be performed twice, once with the carbon dioxide test tube on the bottom; once with the carbon dioxide test tube on the top. You should be able to observe diffusion in both cases, even though carbon dioxide is heavier than air.

Demonstration activities

1. Hydrogen chloride and ammonia – see Student Book page 20. Pay attention to safety – this demonstration needs to be set up very carefully. The concentrated hydrochloric acid and concentrated ammonia solution must be in separate small beakers and kept a good distance apart. Soak one cotton wool ball in each solution and squeeze to remove excess liquid. Use tongs or tweezers to carefully place the cotton wool in the ends of the long tube. Replace the rubber bungs.

2. Brownian motion in a smoke cell. This is not easy to see using a microscope, although using an attached projector will help. There are a number of simulations commercially available which might also help.

Students can then tackle the Developing Investigative Skills section on page 20 of the Student Book.

Note:

For the reaction of the metal powder with copper oxide:

- Wear goggle or a face shield.
- Use safety screens.
- The room should be well ventilated to remove the smoke.
- Protect the bench with hard-board or heat-proof mats.
- Students should stand towards the back of the room.
- This is best preformed at the end of a lesson after which the class leaves.

Instructions:

- Weigh 0.5 g of each reagent on paper.
- Mix by passing them from one piece of paper to the other ad place the mixture in a bottle top (the plastic insert should be burnt out first).
- Place the bottle top on a pipe-clay triangle supported by a tripod.
- Keep heating strongly with a Bunsen burner, retire to where students are standing and watch carefully; the induction period may be a few seconds or as long as a minute.

If the reaction fails to work (probably the magnesium powder has oxidised). DO NOT RETURN TO THE APPARATUS AND POKE THE MIXTURE. It is best to turn the gas off at the main control for the laboratory and allow the mixture to cool before adding to water. Place used bottle tops in the waste.

Technician's notes

You will need the following resources for a demonstration and a class practical.

Be sure to check the latest safety notes on these resources before proceeding.

potassium manganate(VII) crystals and petri dishes
carbon dioxide generator (marble chips and dilute hydrochloric acid), test tubes.
lime water and droppers
long glass tube, cotton wool balls, rubber bungs, stands and clamps (Student book page 20)
concentrated ammonia solution, beaker and tongs
1M hydrochloric acid, beaker and tongs
smoke cell, microscope and dropper

Answers

Developing investigative skills, page 20

1. Using tongs and with the students wearing eye protection and rubber gloves.
2. Avoiding breathing the fumes or use a fume cupboard.
3. The ammonia.
4. It travelled approximately twice as far as the hydrogen chloride.
5. The hydrogen chloride has approximately double the mass of the ammonia.

Activity C1b.3 Elements, atoms, molecules, compounds and mixtures

Learning objectives

- To provide definitions for the terms
- To provide opportunities to test understanding of the terms

Learning outcomes

- Be able to define the terms element, atom, molecule, compound and mixture
- Understand the differences between elements, compounds and mixtures

Common misconceptions

Some students think that all compounds are made up of molecules (that is, they are unaware of the nature of ionic compounds).

Resources

Student Book pages 21–22

Worksheet C1b.3a Elements, atoms, molecules, compounds and mixtures

Files on CD–ROM: C1b.3a_worksheet

Approach

Review the definitions produced by the students in Activity C1b.1 and then provide the definitions you want them to use in the future.

Use Worksheet C1b.3a to check on the students' understanding of the definitions. Refer to the Science in Context box on atoms and molecules on page 22 of the Student Book.

Answers

Page 21

1. Diffusion is the random mixing and moving of particles in liquids and gases.
2. a) Brownian motion was first observed in the movement of pollen grains on water.
   b) The smoke particles are being constantly bombarded by rapidly moving air particles.
3. An element is made up of one kind of atom.

C1b.3a Elements, atoms, molecules, compounds and mixtures

A. Draw lines to link the name to the correct definition.

NAME	DEFINITION
Element	The smallest part of an element that can exist on its own
Atom	A substance containing two or more elements combined together
Molecule	Different substances but not chemically joined in any way
Compound	A substance that cannot be broken down into simpler substances by chemical methods
Mixture	A combination of atoms

B. Complete the table by inserting a Y for Yes when the substance matches the category. (The first one has been done for you.)

Substance	Atom	Molecule	Element	Compound	Mixture
Copper (Cu)	Y		Y		
Pure water (H_2O)					
Dilute hydrochloric acid (HCl/H_2O)					
Oxygen gas (O_2)					
Ammonia gas(NH_3)					

Activity C1b.4 Separation of mixtures

Learning objective

- To reinforce understanding of the different methods of separating mixtures

Learning outcomes

- Be able to describe techniques for separating mixtures including:
    - Simple distillation to separate a solid and liquid from a solution
    - Fractional distillation to separate liquids with different boiling points
    - Filtration to separate a solid from a liquid
    - Crystallisation to obtain a solid from a solution
    - Chromatography to separate different soluble solids from a solution

Common misconceptions

Some students think that these processes involve chemical changes.

Resources

Student Book pages 23–25

Worksheet C1b.4a Separation of mixtures (1)

Worksheet C1b.4b Separation of mixtures (2)

Files on CD–ROM: C1b.4a_worksheet, C1b.4b_worksheet, C1b.4a_tech_notes

Resources for a class practical (see below)

Approach

Experience from Activity C1b.1 will determine whether it is appropriate to provide some practical opportunities for the students relating to the separation of mixtures. If it is, then simple distillation and chromatography lend themselves to class practical work.

Worksheets C1b.4a and C1b.4b can be used to support this work.

Technician's notes

You will need the following resources for a class practical.

Be sure to check the latest safety notes on these resources before proceeding.

a dilute solution of washable blue ink
simple distillation apparatus as shown in Student Book page 23 with associated stands, clamps and Bunsen burners
eye protection
paper chromatography apparatus as shown in Student Book page 25
4 samples of black (water soluble) ink labelled A, B, C and D – two of the inks are the same

Answers

Page 25

1. Less liquid will be lost as vapour/the condenser is much more efficient at condensing the vapour as it is constantly kept cool.

2. B – boiling points.

3. The blue dye is the most soluble (it has moved the furthest up the filter paper).

C1b.4a Separation of mixtures (1)

Simple distillation

This experiment allows you to separate some of the components of ink.

Apparatus

Bunsen burner
blue ink
boiling tube with bung
delivery tube
boiling tube
beaker
clamp and stands
eye protection
porcelain pieces

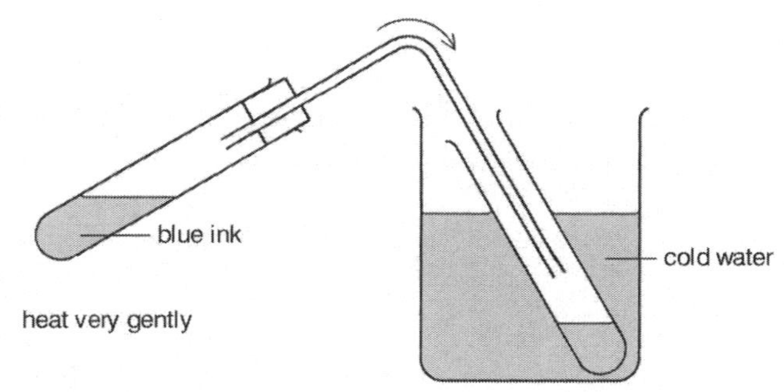

SAFETY INFORMATION

Wear eye protection.

Method

1. Put on eye protection.
2. Set up the apparatus as shown in the diagram, adding a few porcelain pieces to the blue ink in the boiling tube.
3. Heat the ink slowly with a Bunsen burner until you have collected a 1 cm depth of liquid in the test tube.

Demonstrate and describe techniques

1. What was the purpose of adding the porcelain pieces to the ink before heating?
2. What is the function of the cold water in the beaker?
3. What is the general name given to liquids like the one collected in the second test tube?

Analyse and interpret data

1. What evidence is there to suggest that the ink was a mixture?

Evaluate data and methods

1. How could the efficiency of this separation of the components in ink be improved?

C1b.4b Separation of mixtures (2)

Paper chromatography

This experiment allows you to identify which two of the four samples of black ink (A, B, C and D) are the same.

Apparatus

beaker

strip of filter paper

four samples of black ink

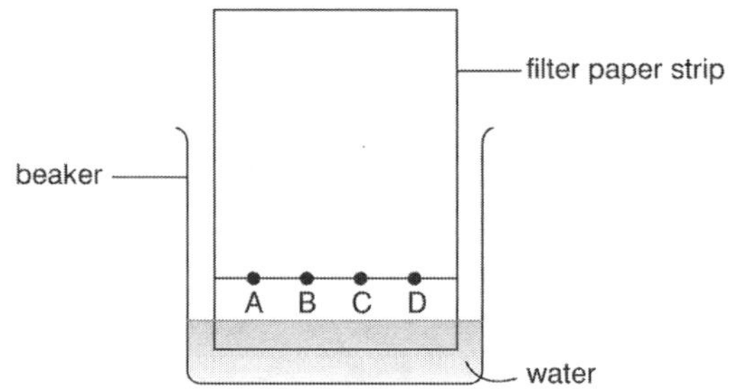

Method

1. Draw a pencil line about 2 cm from the bottom of the filter paper strip.
2. In pencil on this line mark the positions (A, B, C and D) where the four samples of ink will be 'spotted'.
3. Use a different dropper or glass rod for each ink. Add a small drop of each ink onto the mark made on the paper for that ink.
4. Put a 1 cm depth of water in a beaker and carefully place the filter paper into the beaker.
5. When the water has soaked three-quarters of the way up the filter paper strip, remove the strip from the water and let it dry on a paper towel.

Demonstrate and describe techniques

1. Why was a pencil used to draw the line and label the positions of the ink on the filter paper strip?
2. Why was only 1cm depth of water put into the beaker?
3. Why was the filter paper strip paper removed when the water was soaking up close to the top of the paper?

Analyse and interpret data

4. Which coloured dye was the most soluble in water?
5. Which two inks were the same? Explain how you came to your decision.

Activity C1b.5 Consolidation and summary

Learning objectives

- To review the key learning points of the topic
- To test understanding through answering questions

Learning outcomes

- Be familiar with the knowledge and understanding summarised in the End of Topic Checklist
- Be able to apply this knowledge and understanding by answering the End of Topic Questions

Resources

Student Book pages 26–27

Approach

Focus on the End of Topic Checklist and Questions. Aim for a blend of individual and group work and give priority to listening to the students' ideas and explanations.

Answers

End of Topic Questions mark scheme

Question	Correct answer	Marks
1	Diffusion	1 mark
2	There is no physical barrier between the particles of bromine and air. The particles in these gases are moving at high speed and colliding with each other. Over time the mixing will be complete and so there will be air and bromine particles in both gas jars.	1 mark 1 mark
3 a)	You will see small specks of light moving at random.	1 mark
3 b)	The smoke particles reflect the light and are being knocked about by rapidly moving air particles.	1 mark
4	An atom is a single particle whereas a molecule is made up from two or more atoms.	1 mark
5	A compound is a fixed composition/combination of elements joined together; a mixture has no fixed composition.	1 mark
6 a)	Chromatography	1 mark
6 b)	Filtration	1 mark
6 c)	Fractional distillation	1 mark
6 d)	Evaporation or distillation	1 mark
7	A distillate is a liquid that has been evaporated and then condensed (as in a distillation experiment).	1 mark
8	The inks in the black ink do not dissolve in water/are not soluble in water. Using a different solvent such as alcohol/ethanol (this solvent might dissolve some of the dyes in the ink and so they would move up the paper).	1 mark 1 mark
	Total:	14 marks

C1c Atomic structure

Introduction

This topic is the basis for future work on chemical bonding and the Periodic Table. It also introduces relative atomic masses and the concept of a mole of atoms. These quantitative aspects will be further developed to use in 'reacting masses' calculations. The work on atomic structure introduces the sub-atomic particles, leading to drawing and interpreting diagrams showing the numbers of protons and neutrons and arrangement of electrons in shells. There are no opportunities for practical work but plenty of opportunities for individual and group work.

Links to other topics

Sections	Essential background knowledge	Useful links
1 Principles of chemistry	1b Atoms	1d Relative formula masses and molar volumes of gases 1e Chemical formulae and chemical equations 1f Ionic compounds 1g Covalent substances
2 Chemistry of the elements		2a The Periodic Table

Topic overview

C1c.1	**Orientation** Students may have some knowledge of atomic structure and of particles called protons, neutrons and electrons. They may know something about how elements are arranged in the Periodic Table. The purpose of this activity is to gauge their level of understanding to decide how much teaching time to spend on the topic.
C1c.2	**Sub-atomic particles** This activity focuses on the protons, neutrons and electrons in an atom and how the standard representation of atomic number and mass number gives information about the number of these particles in an atom. The arrangement of electrons is covered in Activity C1c.4. You could combine the two activities into one.
C1c.3	**Relative atomic mass and the mole** This activity may be students' first introduction to quantitative chemistry. The relative atomic mass scale is introduced with a mole of atoms. Simple calculations show the conversion of moles of atoms into the mass of atoms. The calculation of the 'average' relative atomic mass of isotopic elements is also covered.
C1c.4	**Electron arrangements** The arrangement of electrons in shells around the nucleus is introduced and the use of atom diagrams to represent the structure of an atom. The significance of the number of electrons in the outer electron shell is emphasised, as is the relationship between that and the position of an element in the Periodic Table. This provides a very useful introduction to Topic 2a The Periodic Table.
C1c.5	**Consolidation and summary** This activity provides a quick recap of the key ideas of the topic.

Activity C1c.1 Orientation

Learning objectives

- To revise students' knowledge about atoms
- To review students' existing knowledge of atomic structure

Learning outcomes

- Know that atoms are the building blocks of all materials
- Understand that atoms are made up of smaller sub-atomic particles which give the atom its characteristic properties

Common misconceptions

Some students think that all atoms of the same element are the same (that is, they ignore the existence of isotopes).

Resources

Student Book page 28

Worksheet C1c.1a What do you know about atomic structure?

Files on CD–ROM: C1c.1a_worksheet

Approach

The orientation exercise can be based around the True/False quiz on Worksheet C1c.1a. The 'not sure' column is meant to stop students from guessing. You can then share the overall objectives of the topic and refer to the quantitative aspects. This is important because it opens up a way for chemists to calculate *how much* of a particular chemical can be made in a reaction.

Students often have problems with the concept of moles. It is important for you to emphasise that it will be covered several times throughout the course and so students can develop a complete understanding over time.

C1c.1a What do you know about atomic structure?

This quiz is intended to find out what you already know about atoms and their structure. For each statement, choose one option from TRUE, FALSE, NOT SURE.

STATEMENT	TRUE	FALSE	NOT SURE
1. All substances are made up of atoms.			
2. An atom is the smallest particle of an element that can exist on its own.			
3. Hydrogen is the smallest/ lightest atom.			
4. Atoms are always made up of three sub-atomic particles called protons, neutrons and electrons.			
5. Protons are the smallest of these sub-atomic particles.			
6. The sub-atomic particles are found in the nucleus of the atom.			
7. A proton is a particle with a negative charge.			
8. A neutron is a particle with no charge.			
9. Isotopes are atoms of the same element with different numbers of neutrons.			
10. Isotopes are radioactive.			
11. Elements are arranged in the Periodic Table in order of the number of protons they have.			
12. Elements are arranged in the Periodic Table in order of the number of electrons they have.			

Activity C1c.2 Sub-atomic particles

Learning objectives

- To introduce the structure of the atom
- To show how atomic structure can be presented in terms of atom diagrams

Learning outcomes

- Know that an atom is made up of a central nucleus, composed of protons and neutrons, surrounded by electrons arranged in shells
- Know the relative masses and charges of protons, neutrons and electrons
- Understand the meaning of the terms atomic number, mass number and isotope

Common misconceptions

Some students think that all isotopes are radioactive.

Resources

Student Book pages 28–31

Approach

This is a relatively brief activity and all the key learning points are covered in the Student Book pages. Electron arrangements and atom diagrams are covered in Activity C1c.4, unless you have decided to combine these two activities.

Once you have introduced the relative masses and charges of the proton, neutron and electron and the atomic number/mass number notation, ask students to use this notation to work out the structures of a number of atoms. Some possible examples include:

$^{56}_{26}Fe$ $^{127}_{53}I$ $^{108}_{47}Ag$ $^{197}_{79}Au$ $^{80}_{35}Br$ $^{35}_{17}Cl$ $^{37}_{17}Cl$

Chlorine is included to introduce the idea of isotopes. Looking at the Periodic Table shows that the atomic number is used but the relative atomic mass is used in preference to the mass number. For a number of elements, the relative atomic mass is not whole number (for example, Cu 63.5) – this clearly shows that the atom must have a number of different isotopes. The concept of relative atomic mass is covered in more detail in the next activity.

Answers

Page 30

1. The electron has the smallest relative atomic mass.

2. Atoms have no overall charge, so the number of positive charges (protons) and negative charges (electrons) must cancel out.

Activity C1c.3 Relative atomic mass and the mole

Learning objectives

- To introduce the relative mass scale for comparing the mass of different atoms
- To introduce the concept of a mole of atoms

Learning outcomes

- Know that the relative atomic mass of an atom is the relative mass of an atom on a scale where the mass of a carbon-12 atom is 12
- Understand that the mole is a number equal to 6×10^{23}
- Be able to perform simple calculations relating moles, mass and relative atomic mass
- Be able to calculate the relative atomic mass of an element from the relative abundances of its isotopes

Common misconceptions

Some students think that the relative mass scale is based on a comparison of atom masses with that of the hydrogen atom (that is, as it used to be).

Resources

Student Book pages 32–34

Worksheet C1c.3a Moles of atoms

Files on CD–ROM: C1c.3a_worksheet

Approach

Explain the need for a relative atomic mass scale because atoms are too small to weigh, and explain that the original standard was hydrogen but that it is now the carbon-12 isotope.

Use the Student Book page 32 to illustrate this relative mass scale. Explain that if these relative atomic masses are expressed in grams then each element must contain the same number of atoms: that is, the Avogadro number or constant (the mole).

Discuss the worked examples in the Student Book and then use Worksheet C1c.3a to let students practise manipulating the conversion of mass to moles and vice versa. Finally, explain the calculation of 'average' relative atomic masses of elements with isotopes, using the example of chlorine on page 33 of the Student Book.

Complete the activity by asking students to answer the questions on page 34 of the Student Book.

Answers

Page 34

1. Isotopes are atoms of the same element with different atomic masses (different numbers of neutrons).

2. 96 g

3. $A_r = \dfrac{(51\times 79)+(49\times 81)}{100} = 79.98$

C1c.3a Moles of atoms

Use the following formula to answer the questions:

$$\text{Moles} = \frac{\text{mass}}{\text{relative atomic mass}}$$

The A_r values are given in brackets.

1. What is the mass of the following?
   a) 1 mole of carbon atoms (C = 12)
   b) 3 moles of sodium atoms (Na = 23)
   c) 0.1 moles of magnesium atoms (Mg = 24)
   d) 2 moles of sulfur atoms (S = 32)
   e) 0.5 moles of calcium atoms (Ca = 40)
   f) 0.01 moles of silicon atoms (Si = 28)

2. How many moles of atoms are there in the following?
   a) 54 g of aluminium (Al = 27)
   b) 4 g of argon atoms (Ar = 40)
   c) 36 g of beryllium atoms (Be = 36)
   d) 71 g of chlorine atoms (Cl = 35.5)
   e) 0.635 g of copper atoms (Cu = 63.5)

3. What are the relative atomic masses of the following? Which element is each one?
   a) 3 moles of atoms has a mass of 3 g
   b) 0.5 moles of atoms has a mass of 40 g
   c) 0.1 moles of atoms has a mass of 3.2 g
   d) 0.25 moles of atoms has a mass of 10 g
   e) 0.01 moles of atoms has a mass of 0.56 g

Activity C1c.4 Electron arrangements

Learning objectives

- To explain how electrons are arranged in shells and how the number in the outer shell relates to the element's position in the Periodic Table
- To introduce atom diagrams as a means of representing the atomic structure of atoms

Learning outcomes

- Understand that the Periodic Table is an arrangement of elements in order of their atomic numbers
- Know that the arrangement of electrons in an atom can be worked out from its position in the Periodic Table
- Know that the number given to a group of elements in the Periodic Table is the same as the number of electrons in the outer electron shell of the atoms in the group

Resources

Student Book pages 34–37

Worksheet C1c.4a Atomic structures

Files on CD–ROM: C1c.4a_worksheet

Approach

This activity follows closely from Activity C1c.2 and you may have decided to combine them. Introduce the idea that electron shells can hold a maximum number of electrons, and that these numbers differ depending on which shell is being occupied by electrons.

Use the Student Book to illustrate the nature of atom diagrams and then the detailed electronic structure of the first 20 elements. You can then make the link between the number of outer electrons and the position of the element in the Periodic Table. This idea will be revised at the start of the Topic 2a *The Periodic Table*.

Use Worksheet C1c.4a to reinforce the ideas introduced in this activity.

Answers

Page 37

1. 2 electrons. Magnesium is in Group 2.

2. a) Aluminium b) Calcium

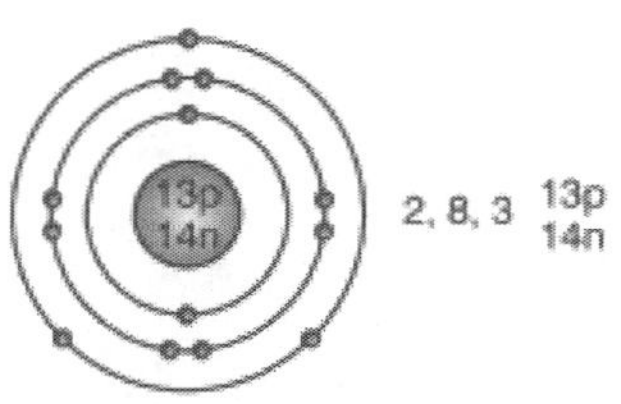

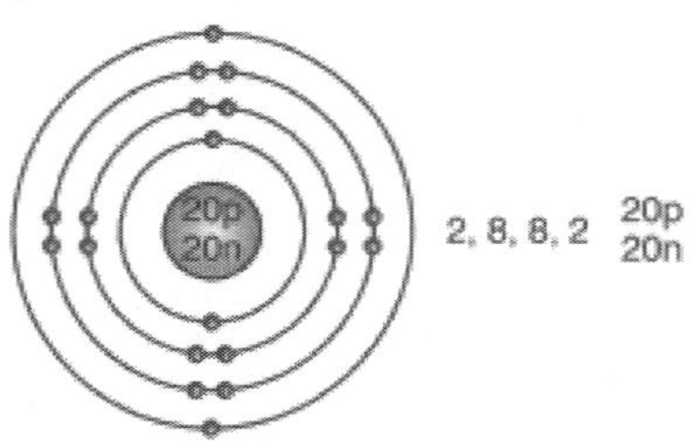

3. The noble gases have a full outer shell of electrons or have 8 electrons in their outer shell and so do not easily lose or gain electrons.

Worksheet C1c.4a

1.

Atom	Atomic number	Mass number	No. of protons	No. of neutrons	No. of electrons
carbon, C	6	12	**6**	**6**	**6**
nitrogen, N	7	**14**	**7**	7	**7**
chlorine, Cl	**17**	35	17	**18**	**17**
sodium, Na	**11**	23	**11**	12	**11**
argon, Ar	**18**	40	**18**	**22**	18
uranium, U	92	238	**92**	**146**	**92**

2. first shell: 2, second shell: 8, third shell: 8

Students' diagrams should have the number of electrons in each shell corresponding to the electronic configurations in Question 3.

3.

element	electronic configuration	element	electronic configuration	element	electronic configuration
hydrogen	1				
helium	2				
lithium	2,1	sodium	**2,8,1**	potassium	**2,8,8,1**
beryllium	**2,2**	magnesium	**2,8,2**	calcium	2,8,8,2
boron	**2,3**	aluminium	**2,8,3**		
carbon	2,4	silicon	**2,8,4**		
nitrogen	**2,5**	phosphorus	**2,8,5**		
oxygen	**2.6**	sulphur	**2,8,6**		
fluorine	2,7	chlorine	**2,8,7**		
neon	**2,8**	argon	2,8,8		

C1c.4a Atomic structures

1. Complete this table for atomic number and mass number.

Atom	Atomic number	Mass number	No. of protons	No. of neutrons	No. of electrons
carbon, C	6	12			
nitrogen, N	7			7	
chlorine, Cl		35	17		
sodium, Na		23		12	
argon, Ar		40			18
uranium, U	92	238			

2. Electrons moving around the nucleus are arranged in shells. How many electrons can each shell hold? Complete this diagram and explain what is meant by an atom's **electronic configuration**.

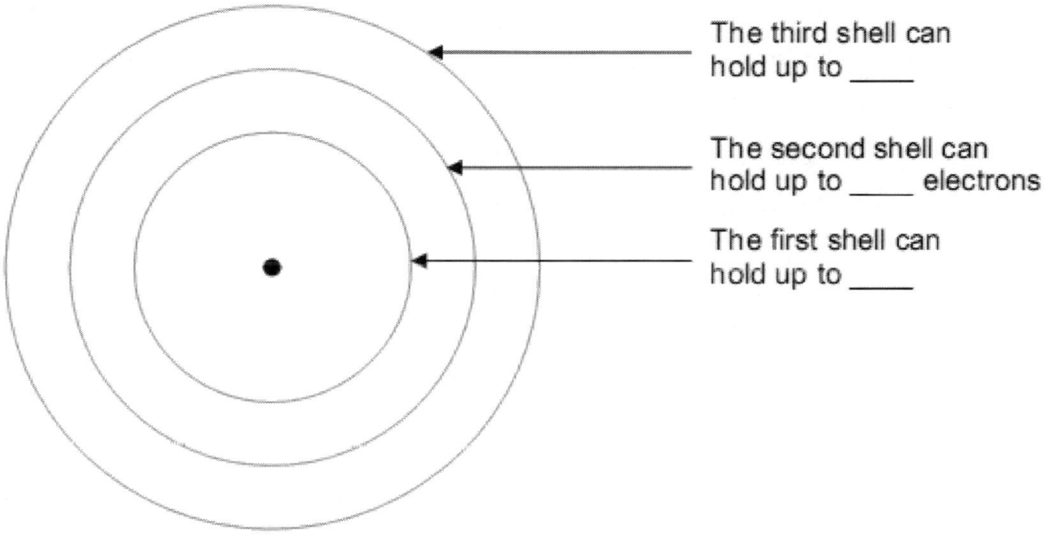

Page 1 of 2

Complete these atom diagrams to show:

a) the composition of the nucleus

b) the electronic configuration, by drawing crosses to represent electrons.

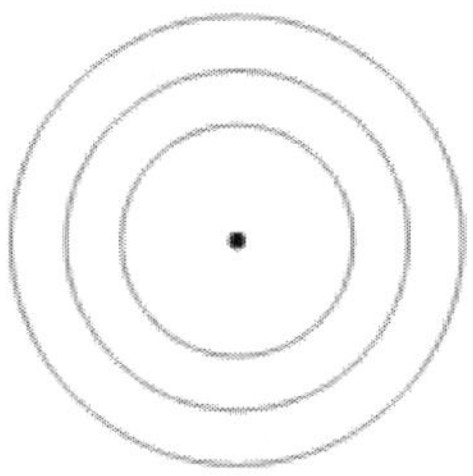	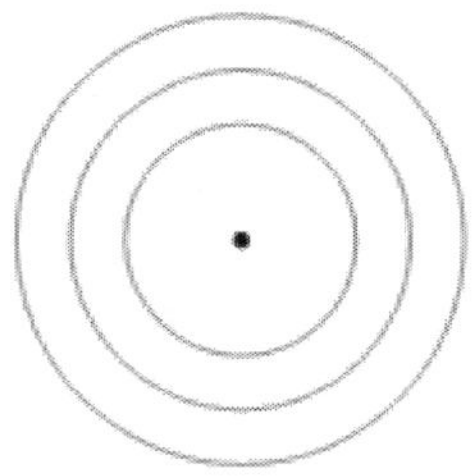	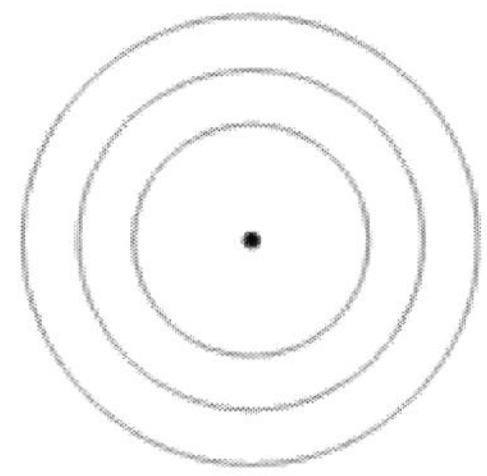	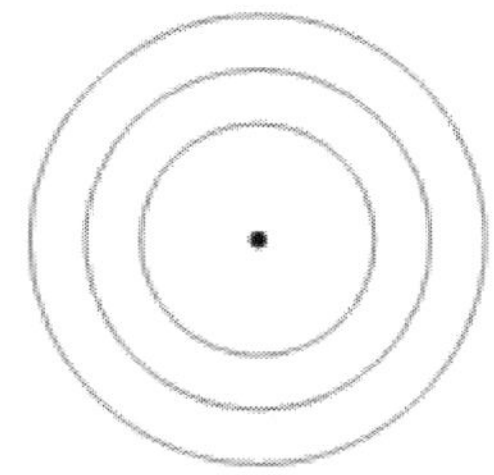
lithium, Li atomic number 3 mass number 7	carbon, C atomic number 6 mass number 12	nitrogen, N atomic number 7 mass number 14	oxygen, O atomic number 8 mass number 16
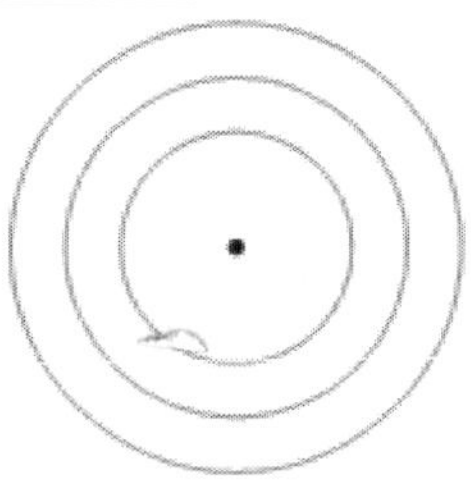	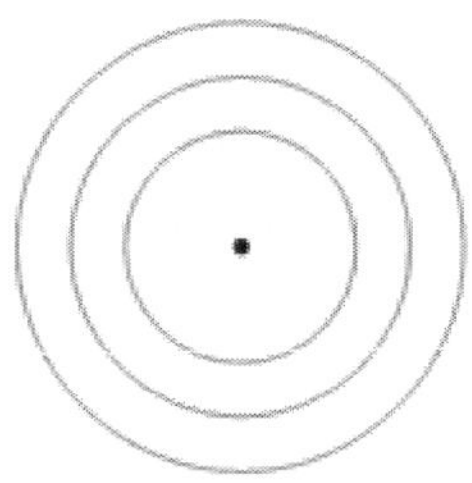	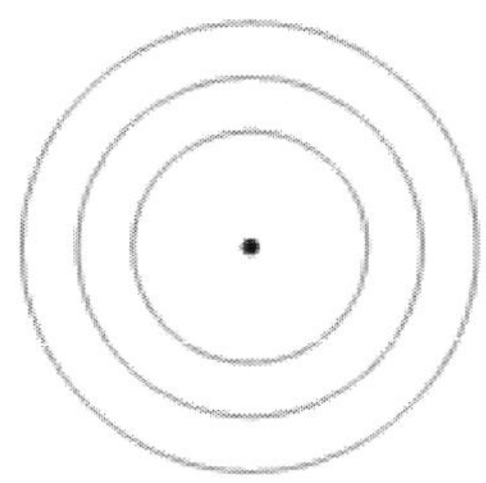	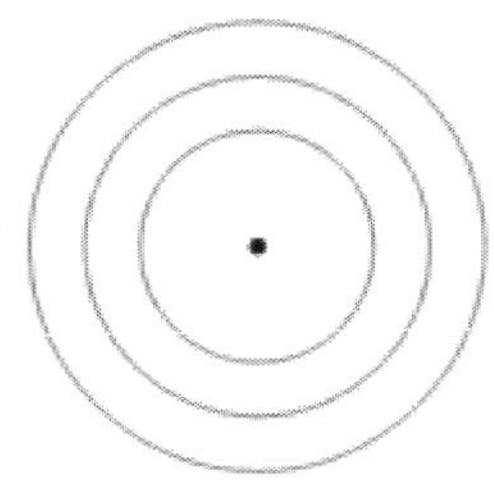
neon, Ne atomic number 10 mass number 20	sodium, Na atomic number 11 mass number 23	chlorine, Cl atomic number 17 mass number 35	argon, Ar atomic number 18 mass number 40

3. a) Complete the missing electronic configurations for the first 20 elements:

element	electronic configuration	element	electronic configuration	element	electronic configuration
hydrogen	1				
helium	2				
lithium	2,1	sodium		potassium	
beryllium		magnesium		calcium	2,8,8,2
boron		aluminium			
carbon	2,4	silicon			
nitrogen		phosphorus			
oxygen		sulphur			
fluorine	2,7	chlorine			
neon		argon	2,8,8		

b) In your own words, say why the electronic configurations of atoms are important.

Page 2 of 2

Activity C1c.5 Consolidation and summary

Learning objectives

- To review the learning points of the topic
- To test understanding through answering questions

Learning outcomes

- Be familiar with the knowledge and understanding summarised in the End of Topic Checklist
- Be able to apply this knowledge and understanding by answering the End of Topic Questions

Resources

Student Book pages 38–40

Approach

Focus on the End of Topic Checklist and Questions. Aim for a mix of individual and group work and provide plenty of opportunities for the students to articulate their ideas and understanding.

Answers

End of Topic Questions mark scheme

Question	Correct answer	Marks
1	A proton has a relative mass of 1 unit.	1 mark
2 a)	The atomic number is the number of protons (= number of electrons) in an atom.	1 mark
2 b)	The mass number is the number of protons + neutrons in an atom	1 mark
3 a)	Isotopes are atoms of the same element with different numbers of neutrons.	1 mark
3 b)	chlorine-35: 17 protons; 18 neutrons; 17 electrons chlorine-37: 17 protons; 20 neutrons; 17 electrons	1 mark 1 mark
4 a)	56/28 = 2 moles	1 mark
4 b)	3.1/31 = 0.1 moles	1 mark
5	Completed table as shown below	4 marks

Atom	Number of protons	Number of neutrons	Number of electrons	Electron arrangement
Si	14	14	14	2,8,4
Mg	12	12	12	2,8,2
S	16	16	16	2,8,6
Ar	18	22	18	2,8,8

Question	**Correct answer**	**Marks**
6 a)	i) C ii) B iii) C or E or F iv) B v) C	1 mark 1 mark 1 mark 1 mark 1 mark
6 b)	Diagram should have 6 protons and 6 neutrons in the nucleus as shown below. 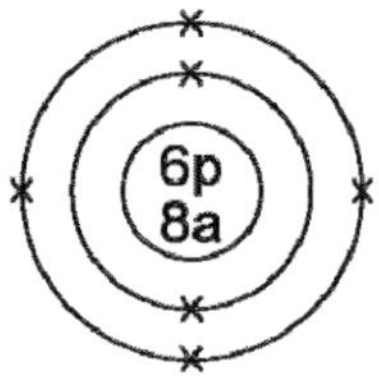	2 marks
7 a)	Diagram should have 8 protons and 8 neutrons in the nucleus as shown below. 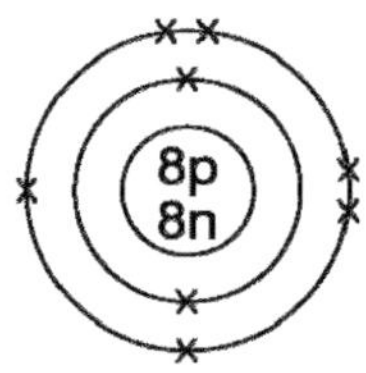	2 marks
7 b)	Diagram should have 19 protons and 20 neutrons in the nucleus as shown below. 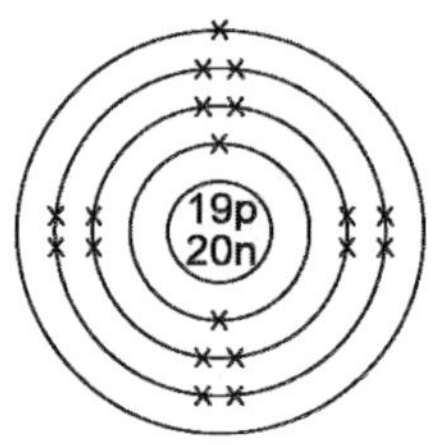	2 marks
8 a)	No	1 mark
8 b)	No	1 mark
8 c)	No	1 mark
8 d)	Yes. The number of electrons in the outer shell is the same as the group number.	1 mark
	Total:	27 marks

C1d Relative formula masses and molar volumes of gases

Introduction

This topic extends the work on relative atomic masses from Topic 1c and provides a direct link to Topic 1e. It introduces two key concepts: relative formula mass and the molar volume of gases.

Although this is a relatively short topic, some students will find the concepts quite difficult to grasp. It is important for you to offer plenty of opportunities for students to apply the ideas and to answer questions. The last activity provides some useful practical work.

Links to other topics

Sections	Essential background knowledge	Useful links
1 Principles of chemistry	1a States of matter 1c Atomic structure	1e Chemical formulae and chemical equations 1i Electrolysis
4 Physical chemistry		4b Energetics

Topic overview

C1d.1	**Orientation** This activity revises ideas on relative atomic masses and the mole before providing an overview of what will be covered in the topic. Relative formula mass is introduced as a way of extending the mole concept to include compounds. Then the molar volume of gases is introduced as a more convenient way of measuring the quantity of gas rather than finding its mass.
C1d.2	**Relative formula masses** This activity focuses on the use of relative atomic masses to calculate relative formula masses and the extension of the mole concept to apply to compounds. It largely involves giving the students plenty of practice with calculations drawing initially on the examples given in the Student Book.
C1d.3	**Molar volumes of gases** A short activity which introduces the notion that 1 mole of any gas will occupy the same volume under the same conditions of temperature and pressure. 24,000 cm^3 is introduced as the molar volume of any gas at room temperature and pressure (RTP). Some worked examples are included in the Student Book to illustrate how the molar volume concept can be used in calculations. This is a brief taste of a more detailed consideration in Topic C1e *Chemical formulae and chemical equations*.
C1d.4	**Consolidation and summary** This activity provides a quick recap of the key ideas of the topic.

Activity C1d.1 Orientation

Learning objectives

- To revise ideas on relative atomic masses and the mole
- To map out the main purposes of the topic

Learning outcomes

- Know that the masses of different elements or atoms are measured on a relative scale
- Understand what relative atomic mass (A_r) is
- Understand that the mole is a number of atoms

Common misconceptions

Some students think that all compounds contain molecules (that is, relative formula mass is the same as relative molecular mass in all cases).

Resources

Student Book pages 41–42

Files on CD–ROM: C1d.1a_tech_notes

Resources for demonstration (see below)

Approach

Summarise the work you previously covered in Topic 1c *Atomic structure:* relative atomic mass, the mole as a number of atoms and simple mole calculations relating to elements.

Explain that elements usually exist in compounds (combined with other elements) and so it is useful to apply the concept of moles to compounds as well as to elements. The first part of the topic deals with this aspect. The second part focuses on expressing molar quantities for gases as volumes, not as masses – it is much more convenient this way.

Demonstrate the reaction of magnesium ribbon and dilute hydrochloric acid, collecting the hydrogen in a gas syringe. Make the point that it is relatively easy to work out how much gas has been produced, but that weighing the gas accurately is very difficult in a school laboratory.

Technician's notes

You will need the following resources for a demonstration.

Be sure to check the latest safety notes on these resources before proceeding.

conical flask, delivery tube with connection to a 100cm^3 gas syringe.
an 8 cm length of magnesium ribbon
25 cm^3 of 1M hydrochloric acid

Activity C1d.2 Relative formula masses

Learning objectives

- To introduce the concept of formula mass and provide practice in calculating formula masses
- To extend the mole concept to include compounds

Learning outcomes

- Know how to calculate relative formula masses from relative atomic masses
- Understand the use of the term mole to represent the amount of a substance
- Know that the Avogadro number of particles is 6×10^{23}

Common misconceptions

Some students think that all compounds contain molecules (that is, relative formula mass is the same as relative molecular mass in all cases)

Resources

Student Book pages 41–43

Worksheet C1d.2a (Q1) Relative formula masses and molar volumes

Files on CD–ROM: C1d.2a_worksheet

Approach

Refer to page 42 of the Student Book for the worked examples showing how to calculate relative formula masses. Emphasise what a radical such as NO_3 represents (1 N atom, 3 O atoms) and how brackets should be interpreted in chemical formulae, as in $Ca(OH)_2$ for example.

Ask students to complete the questions on page 42 of the Student Book. Additional questions are included on Worksheet C1d.2a (Q1).

Finally, extend the application of the mole concept to include relative formula masses.

Answers

Page 42

1. 16
2. 46
3. 48

C1d.2a Relative formula masses and molar volumes

1. Answering these questions will help you to become familiar with working out relative formula masses and molar masses.

a) Calculate the formula masses of the following compounds:

A$_r$: H = 1; C= 12; N = 14; O = 16; Mg = 24; Al = 27; S = 32; Ca = 40; Cu = 64

calcium oxide: CaO __________

propane: C_3H_8 __________

carbon dioxide: CO_2 __________

magnesium sulfate: $MgSO_4$ __________

copper(II) carbonate: $CuCO_3$ __________

nitric acid: HNO_3 __________

aluminium hydroxide: $Al(OH)_3$ __________

copper(II) nitrate: $Cu(NO_3)_2$ __________

b) Calculate the mass of the following:

2 moles of propane __________

0.5 moles of carbon dioxide __________

0.1 moles of nitric acid __________

10 moles of calcium oxide __________

2. Answering these questions will help you to become familiar with working with molar volumes of gases.

1 mole of any gas occupies 24 000 cm^3 at room temperature and pressure (RTP)

Calculate the mass of the following:

48 000 cm^3 of hydrogen: H_2 __________

12 000 cm^3 of carbon dioxide: CO_2 __________

6000 cm^3 of propane: C_3H_8 __________

100 cm^3 of hydrogen: H_2 __________

400 cm^3 of carbon dioxide: CO_2 __________

40 cm^3 of oxygen: O_2 __________

Activity C1d.3 Molar volumes of gases

Learning objectives

- To introduce the concept of the molar volume of gases
- To introduce reacting mass and volume calculations using balanced equations

Learning outcomes

- Know that the molar gas volume for any gas is 24 dm^3 (24 000 cm^3) at room temperature and pressure (RTP)
- Be able to calculate quantities using relative atomic masses, relative formula masses and the mole

Resources

Student Book pages 44–46

Worksheet C1d.2a (Q2) Relative formula masses and molar volumes

Files on CD–ROM: C1d.2a_worksheet

Approach

Introduce 24 000 cm^3 as the molar volume of any gas at RTP. It may seem strange that the actual size of the gas molecules has no effect on the volume occupied by 6×10^{23} of these molecules. This is why scientists didn't accept Avogadro's hypothesis at first (see Student Book page 43). The explanation depends on the fact that the average distance between gas particles is much greater than the diameter of an individual gas particle.

Review the 'reacting mass' calculations in the Student Book to illustrate the use of the molar volume for gases. Use Worksheet C1d.2a (Q2) to reinforce these ideas. There is more detailed discussion of using equations to calculate quantities in chemical reactions in Topic 1e *Chemical formulae and chemical equations*.

Answers

Page 44

1. a) 2 moles

   b) 0.5 moles

   c) 0.1 moles

2. a) 0.5 moles

   b) 0.1 moles

   c) 2 moles

Page 46

1. 1 mole of magnesium (24 g) produces I mole of hydrogen gas (24 000 cm^3). So 1/6 mole of Mg produces 1/6 mole hydrogen = 4000 cm^3 at room temperature and pressure.

2. 1 mole of oxygen gas is produced from 2 moles of hydrogen peroxide. So 0.5 moles of oxygen gas is produced from 1 mole of hydrogen peroxide.

Activity C1d.4 Consolidation and summary

Learning objectives

- To review the learning points of the topic
- To test understanding through answering questions

Learning outcomes

- Be familiar with the knowledge and understanding summarised in the End of Topic Checklist
- Be able to apply this knowledge and understanding by answering the End of Topic Questions

Resources

Student Book pages 47–49

Worksheet C1d.4a Calculating the molar volume of carbon dioxide

Files on CD–ROM: C1d.4a_worksheet, C1d.4a_tech_notes

Resources for a class practical (see below)

Approach

As well as focusing on the End of Topic Checklist and Questions, you can reinforce the ideas with some simple practical work involving the decomposition of copper(II) carbonate. Details are provided on Worksheet C1d.4a.

As well as giving practice in calculating molar volumes, the practical requires careful and accurate quantitative work in order to achieve results of acceptable accuracy.

Technician's notes

You will need the following resources for a class practical.

Be sure to check the latest safety notes on these resources before proceeding.

oven dried test tubes
mineral wool
access to a 2 decimal place balance
copper(II) carbonate
spatulas
tongs
eye protection

Answers

End of Topic Questions mark scheme

Question	Correct answer	Marks
1 a)	28 g	1 mark
1 b)	64 g	1 mark
1 c)	32 g	1 mark
2 a)	64/256 = 0.25 moles	1 mark
2 b)	9.8/98 = 0.1 moles	1 mark
2 c)	21/7 = 3 moles	1 mark
3 a)	2.5 × 88 = 220 g	1 mark
3 b)	0.25 x 40 = 10 g	1 mark
3 c)	0.1 × 109 = 10.9 g	1 mark
4 a)	24 000/24 000 = 1 mole	1 mark
4 b)	1200/24 000 = 0.05 moles	1 mark
5 a)	$Fe_2O_3(s) + 3CO(g) \rightarrow 2Fe(s) + 3CO_2(g)$ 1 mole 2 moles 160 tonnes 112 tonnes 800 tonnes 560 tonnes	 1 mark 1 mark 1 mark
5 b)	$Fe_2O_3(s) + 3CO(g) \rightarrow 2Fe(s) + 3CO_2(g)$ 1 mole 3 moles 160 g 72 000 cm^3 320 g 144 000 cm^3	 1 mark 1 mark 1 mark
6 a)	$2Mg(s) + O_2(g) \rightarrow 2MgO(s)$ 2 moles 1 mole 2 moles 48 g 32 g 80 g 4.8 g 3.2 g 8.0 g	 1 mark 1 mark 1 mark
6 b)	3.2 g	1 mark
	Total:	21 marks

C1d.4a Calculating the molar volume of carbon dioxide

When heated strongly, copper(II) carbonate breaks down into copper(II) oxide and carbon dioxide. Although the carbon dioxide gas will escape during the reaction, the mass of the gas can be calculated from the loss in total mass during the reaction. This loss in mass can then be converted into a corresponding volume.

Copper(II) carbonate → copper(II) oxide + carbon dioxide

$$CuCO_3(s) \rightarrow CuO(s) + CO_2(g)$$

Apparatus

dry test tube and test tube holder

mineral wool

tongs

copper(II) carbonate

spatula

access to a balance weighing to 2 decimal places

eye protection

SAFETY INFORMATION

Copper(II) carbonate is harmful. Avoid direct contact with the powder.

Only handle the mineral wool with tongs. Do NOT touch it with fingers.

Method

1. Put on eye protection.
2. Weigh a clean, dry test tube with a small plug of mineral wool near the top of the tube. Record your result in a table.
3. Remove the mineral wool and carefully add about 2 spatula measures of the copper(II) carbonate powder to the tube. Replace the plug of mineral wool and reweigh the tube. Again, record your result.
4. Heat the tube, slowly at first to prevent the copper(II) carbonate powder from escaping from the tube, and then more strongly until all the green powder has changed to a black powder.
5. Allow the tube to cool in a test tube rack.
6. Once cool, reweigh the tube and contents. Record your result.

Devise and plan investigations

1. Why was the mineral wool used?

Make observations and measurements

1. Record your results in a suitable table.
2. Use your results to calculate the mass of carbon dioxide produced in the reaction.

Analyse and interpret data

1. Calculate the number of moles of carbon dioxide produced in the reaction (A_r: C=12; O = 16).

2. Calculate the volume this number of moles would occupy at RTP (1 mole of gas occupies 24 000 cm^3 at RTP).

3. Calculate the number of moles of copper(II) carbonate used (A_r: C = 12; O = 16, Cu = 64).

4. Use your answer from Question 3, and the equation, to calculate what volume of carbon dioxide should have been produced in the reaction (the theoretical volume according to the equation).

Evaluate data and methods

1. The result you calculated from your experiment and the theoretical result are likely to be different. Suggest why.

C1e Chemical formulae and chemical equations

Introduction

This is a pivotal topic in the course. It introduces chemical equations as the chemist's 'shorthand' for both describing reactions and providing quantitative details. The quantitative work builds on Topics 1c *Atomic structure* (relative atomic mass) and 1d *Relative formula masses and molar volumes of gases*. Ionic equations follow on from Topic 1e *Ionic compounds* and link well to Topic 4a *Acids, alkalis and salts*.

The practical work provides a good opportunity to practise some of the investigative skills, particularly *Make observations and measurements*, *Analyse and interpret data* and *Evaluate data and methods*. Section 5 *Chemistry in society* reinforces the quantitative aspects.

Links to other topics

Sections	Essential background knowledge	Useful links
1 Principles of chemistry	1c Atomic structure 1d Relative formula masses and molar volumes of gases	
2 Chemistry of the elements		2g Tests for ions and gases
4 Physical chemistry		4a Acids, alkalis and salts 4d Equilibria
5 Chemistry in society		5d The industrial manufacture of chemicals

Note: All topics after this one will make use of chemical formulae and chemical equations. The ones listed as useful links should be the strongest links.

Topic overview

C1e.1	**Orientation** The purpose of this activity is to check students' recall and understanding of the ideas encountered so far on relative atomic and formula masses, the mole and ionic bonding.
C1e.2	**Finding the formula of magnesium oxide** This activity involves reacting magnesium ribbon with oxygen from the air using a crucible and lid to minimise the loss of magnesium oxide. The results can be used to work out the mole ratio of magnesium and oxygen and hence the formula of magnesium oxide.
C1e.3	**Empirical and molecular formulae** This activity introduces the ideas of empirical and molecular formulae and illustrates how both types of formulae can be determined.
C1e.4	**Finding the formula of copper(II) oxide** This activity involves the reduction of copper(II) oxide to copper using the small amounts of hydrogen present in natural gas. The activity could follow on immediately after Activity C1e.2. It is presented as the fourth activity to provide some variety of activity through a topic which includes a relatively large amount of work on chemical calculations.

C1e.5	**Writing chemical equations** The students should be familiar with word equations and this activity is intended to help them to become more familiar with symbol equations and how they can be balanced.
C1e.6	**Reacting masses** This activity has two distinct parts. The first part involves a class practical involving the decomposition of zinc carbonate, which illustrates how an equation can be used to check the accuracy of experiment results, as well as introducing the concept of percentage yield. The second part enables the students to practise reacting mass calculations.
C1e.7	**Consolidation and summary** This activity provides a quick recap of the key ideas encountered in the topic.

Activity C1e.1 Orientation

Learning objectives

- To revise ideas on relative atomic masses, formula masses and the mole
- To map out the main purposes of the topic

Learning outcomes

- Know how to work out a formula mass from atomic masses
- Understand that a mole is a number that can be applied to atoms and molecules

Resources

Student Book page 50

Files on CD–ROM: C1e.1a_tech_notes

Resources for a demonstration (see below)

Approach

Briefly summarise the ideas from Topics 1c *Atomic structure* and 1d *Relative formula masses and molar volumes of gases*. Ask students some questions to help with this.

Demonstrate the burning of magnesium ribbon in air. Ask students to explain what is happening and what the white ash/smoke is. Explain that if the mass of oxygen that combines with a certain mass of magnesium can be measured, then the formula of magnesium oxide can be worked out.

CAUTION: Do not look at the flame.

After the demonstration, ask students if there are any particular practical problems with this process. They should point out that magnesium oxide escapes into the atmosphere. Explain that a way must be found to prevent this from happening.

The topic will look in more detail at chemical equations and how they can be balanced. Explain that the topic will also discuss using equations to calculate the quantities of reactants and products needed or made in a chemical reaction.

Technician's notes

You will need the following resources for demonstration purposes.

Be sure to check the latest safety notes on these resources before proceeding.

Magnesium ribbon
Metal tongs

Activity C1e.2 Finding the formula of magnesium oxide

Learning objectives

- To allow students to undertake some exacting quantitative practical work
- To show how a chemical formula can be worked out from combining masses of elements

Learning outcomes

- Understand how the formula of a simple compound such as a metal oxide can be worked out from experimental results
- Be able to make accurate observations and measurements
- Be able to analyse and interpret data to work out the formula of magnesium oxide
- Be able to evaluate data and methods
- Understand how the formulae of water and salts containing water of crystallisation can be worked out from experimental results

Resources

Student Book pages 50–52

Worksheet C1e.2a Working out the formula of magnesium oxide

Files on CD–ROM: C1e.2a_worksheet, C1e.2a_tech_notes

Resources for a class practical (see below)

Approach

The orientation activity reminded students about the reaction between magnesium and oxygen. Explain that they must now undertake this reaction very carefully. They need results that are accurate enough to let them work out the formula of magnesium oxide.

The experiment is described on Worksheet C1e.2a. It is also on pages 50–52 of the Student Book. The Student Book section could be used as an introduction to the experiment or as consolidation after the practical work is completed.

The crucial part of the experiment is to minimise the loss of the magnesium oxide powder. Use the Student Book to explore how similar calculations can be used to find the formulae of copper(II) oxide, water and salts such as hydrated copper(II) sulfate.

Technician's notes

You will need the following resources for a class practical.

Be sure to check the latest safety notes on these resources before proceeding.

crucible and lid
pipe-clay triangle
tripod
Bunsen burner
metal tongs/eye protection
magnesium ribbon
emery cloth
access to a balance weighing up to 2 decimal places

C1e.2a Working out the formula of magnesium oxide

Magnesium burns in air to form magnesium oxide. The formula of magnesium oxide can be worked out if the exact masses of magnesium and oxygen which react together can be worked out.

Apparatus

crucible and lid

emery paper

pipe-clay triangle

tongs

tripod

Bunsen burner

magnesium ribbon

access to a balance, accurate to 2 decimal places

eye protection

SAFETY INFORMATION

Magnesium is HIGHLY FLAMMABLE.
Wear eye protection.

Method

1. Put on eye protection.

2. Weigh a clean, dry crucible and lid (A).

3. Clean a 10 cm strip of magnesium ribbon with emery paper. Make the magnesium into a coil and place in the crucible. Weigh the crucible, lid and magnesium (B).

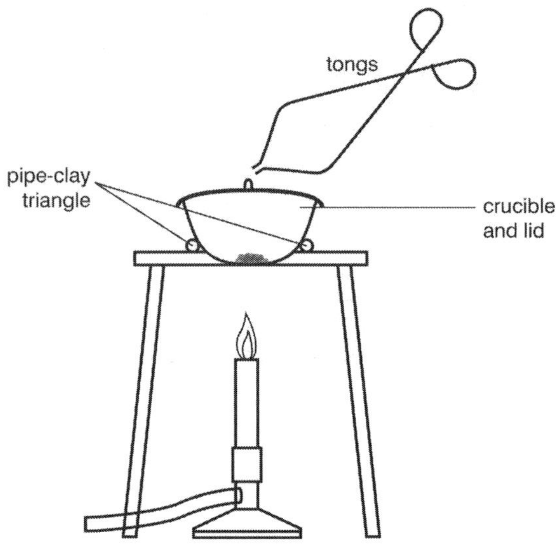

4. Set up the apparatus as shown and heat the crucible strongly. From time to time carefully lift the crucible lid a very small distance and then replace it. TRY TO AVOID ANY LOSS OF WHITE SMOKE.

5. When there is no further sign of a reaction allow the crucible and lid to cool. Then weigh the crucible, lid and magnesium oxide (C).

Make observations and measurements

Record your results in the table below:

Mass of crucible + lid (A) = ____________ g

Mass of crucible + lid + magnesium (B) = ____________ g

Mass of crucible + lid + magnesium oxide (C) = ____________ g

Mass of magnesium (B – A) = ____________ g

Mass of oxygen (C – B) = ____________ g

Analyse and interpret data

A_r O = 16; Mg = 24

$$\text{Number of moles of magnesium} = \frac{\text{mass of magnesium}}{24} = \_\_\_\_\_\_\_\_\_\_\_\_$$

$$\text{Number of moles of oxygen} = \frac{\text{mass of oxygen}}{16} = \_\_\_\_\_\_\_\_\_\_\_\_$$

$$\text{Ratio } (n) = \frac{\text{number of moles of oxygen}}{\text{number of moles of magnesium}} = \_\_\_\_\_\_\_\_\_\_\_\_$$

Formula of magnesium oxide = MgO_n

Evaluate data and methods

1. Why was it important to avoid losing the white smoke?
2. Why was it necessary to clean the magnesium ribbon with emery paper?
3. Magnesium oxide has the formula MgO. What was the percentage error in your experiment?
4. What were the sources of error in the experiment?

Activity C1e.3 Empirical and molecular formulae

Learning objectives

- To introduce the empirical formula as a simple whole number ratio of the atoms in the compound
- To emphasise the difference between an empirical formula and a molecular formula

Learning outcomes

- Be able to calculate empirical and molecular formulae from experimental data

Resources

Student Book pages 54–56

Approach

Explain that the process for determining the chemical formula of magnesium oxide will give the simplest whole number ratio of the atoms. In some cases this will not be the actual molecular formula – hydrogen peroxide is a good example to quote. The Student Book (pages 54–56) includes some worked examples and then some questions for students to answer.

Answers

Page 54

1. a) To allow oxygen from the air into the crucible
   b) To prevent the loss of the magnesium oxide
2. a) Platinum
   b) Oxygen

Page 56

1. Fe_2O_3 (0.10 moles of Fe reacts with 0.15 moles of O)
2. ZnO (0.20 moles of Zn combine with 0.20 moles of O)
3. C_4H_{10} (empirical formula mass is 29 – half of the relative formula mass)
4. H_2O_2 (empirical formula mass is 17 – half of the relative formula mass)

Activity C1e.4 Finding the formula of copper(II) oxide

Learning objectives

- To provide another experimental method for determining the formula of a metal oxide
- To encourage accurate quantitative work

Learning outcomes

- Understand how the formula of copper(II) oxide can be worked out from experimental results
- Be able to make accurate observations and measurements
- Be able to analyse and interpret data
- Be able to evaluate data and methods

Resources

Student Book pages 51–53

Worksheet C1e.4a Working out the formula of copper(II) oxide

Files on CD–ROM: C1e.4a_worksheet, C1e.4a_tech_notes

Resources for a class practical (see below)

Approach

This experiment gives students another opportunity to do some exacting quantitative practical work. They should already have been introduced to the formula of copper(II) oxide in the Student Book. The purpose of this experiment is to try and get accurate results to verify the formula of copper(II) oxide.

The details of the reduction of copper(II) oxide are included on Worksheet C1e.4a. Again, the Student Book includes a worked example of a similar experiment using hydrogen.

In addition, the *Developing Investigative Skills* feature on pages 52–53 of the Student Book relates to the experiment. You will need to decide at what stage to use *Developing Investigative Skills*, as the questions in the Student Book overlap at some points with the worksheet.

Technician's notes

You will need the following resources for a class practical.

Be sure to check the latest safety notes on these resources before proceeding.

rubber tubing
connecting tube and bung
combustion tube (oven dry)
Bunsen burner
stand and clamp
spatula
copper(II) oxide
eye protection

Answers

Developing investigative skills, page 52

1. It will not be possible to heat any copper(II) oxide near the mouth of the tube without melting the rubber bung.
2. To prevent the gas pressure from blowing the copper(II) oxide through the jet.
3. To prevent the hot copper reacting with oxygen in the air.
4. Black (copper(II) oxide) to pink (copper).
5. Mass of copper(II) oxide = 1.00 g

   Mass of copper = 0.77 g

   Mass of oxygen = 0.23 g
6. Number moles of copper = 0.77/64 = 0.012

   Number moles of oxygen= 0.23/16 = 0.014
7. Ratio Cu: O is 1 : 1.17
8. CuO
9. a) loss of copper(II) oxide during weighing or through the jet during heating

   b) not all copper(II) oxide converted into copper

   c) some hot copper reformed copper(II) oxide if the cooling was not complete when the gas flow stopped.

C1e.4a Working out the formula of copper(II) oxide

To work out the formula of copper(II) oxide you must find out the exact masses of copper and oxygen that combine together. In this experiment you start with copper(II) oxide and remove the oxygen to form copper. The removal of oxygen (reduction) is accomplished by heating the copper(II) oxide in a stream of natural gas, which contains a small amount of hydrogen.

Apparatus

rubber tubing

connecting tube and bung

combustion tube

Bunsen burner, stand and clamp

eye protection

SAEFTY INFORMATION

Copper(II) oxide is DANGEROUS.

Wear eye protection.

Method

1. Put on eye protection.
2. Weigh a dry combustion tube (A).
3. Carefully put a spatula measure of copper(II) oxide two-thirds of the way along the tube. Remove any copper(II) oxide that has been spilt near the mouth of the tube. Reweigh the tube and copper(II) oxide (B).
4. Set up the apparatus as shown on the right. Turn on the gas supply very slowly and light the gas at the jet and leave on for 10 s before lighting. The flame should be a maximum of 5 cm high. Heat the copper(II) oxide strongly until it has all turned into pink copper. If the flame at the jet goes out, relight it.
5. Stop heating but leave the gas passing through the tube until the tube has cooled down.

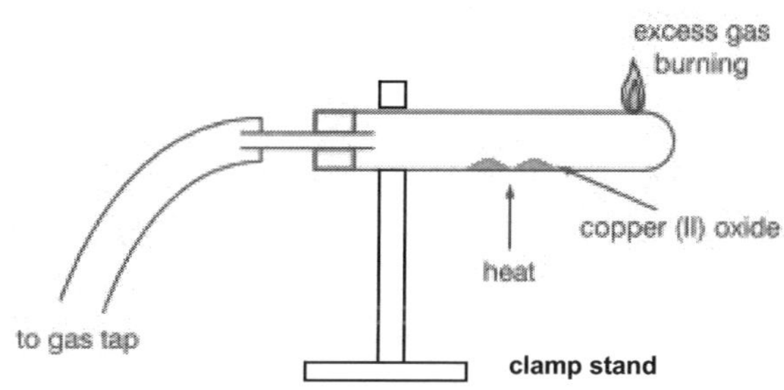

6. When the combustion tube is cool, turn off the gas supply and weigh the tube and the copper (C).

Make observations and measurements

Complete the table of results.

Mass of combustion tube (A)	=	____________ g
Mass of combustion tube + copper(II) oxide (B)	=	____________ g
Mass of combustion tube + copper (C)	=	____________ g
Mass of copper (C – A)	=	____________ g
Mass of oxygen (B – C)	=	____________ g

Page 1 of 2

Analyse and interpret data

O = 16; Cu = 64

Number of moles of copper = $\frac{\text{mass of copper}}{64}$ = ______________

Number of moles of oxygen = $\frac{\text{mass of oxygen}}{16}$ = ______________

Ratio (n) = $\frac{\text{number of moles of oxygen}}{\text{number of moles of copper}}$ = ______________

Formula of copper(II) oxide = CuO_n

Evaluate data and methods

1. Why was it important not to have too great a flow of gas through the tube?
2. Why was it necessary to leave the natural gas flowing over the copper whilst the combustion tube was cooling?
3. Copper(II) oxide has the formula CuO. What was the percentage error in your experiment?
4. What were the sources of error in the experiment?

Activity C1e.5 Writing chemical equations

Learning objectives

- To write fully balanced chemical equations

Learning outcomes

- Be able to write balanced equations, including state symbols, for a range of chemical reactions

Resources

Student Book pages 56–60

Worksheet C1e.5a Balancing equations

Files on CD–ROM: C1e.5a_worksheet

Approach

Encourage students to use state symbols so they get used to doing this consistently. The Student Book provides a basic method for balancing equations.

Worksheet C1e.5a provides examples to try. The Extension question on page 60 of the Student Book includes some more difficult examples which even the best students will find challenging!

Answers

Page 59

1. a) $2Ca(s) + O_2(g) \rightarrow 2CaO(s)$

   b) $2H_2S(g) + 3O_2(g) \rightarrow 2SO_2(g) + 2H_2O(l)$

   c) $2Pb(NO_3)_2(s) \rightarrow 2PbO(s) + 4NO_2(g) + O_2(g)$

2. a) $S(s) + O_2(g) \rightarrow SO_2(g)$

   b) $2Mg(s) + O_2(g) \rightarrow 2MgO(s)$

   c) $CuO(s) + H_2(g) \rightarrow Cu(s) + H_2O(l)$

Page 60

1. a) $2C_5H_{10}(g) + 15O_2(g) \rightarrow 10CO_2(g) + 10H_2O(l)$

   b) $Fe_2O_3(s) + 3CO(g) \rightarrow 2Fe(s) + 3CO_2(g)$

   c) $2KMnO_4(s) + 16HCl(aq) \rightarrow 2KCl(aq) + 2MnCl_2(aq) + 8H_2O(l) + 5Cl_2(g)$

C1e.5a Balancing equations

Balance the following equations. Remember:

- You must not change any chemical formulae.
- You can only put numbers in front of the formulae.

1. $CO(g)$ + $O_2(g)$ → $CO_2(g)$

2. $H_2(g)$ + $Cl_2(g)$ → $HCl(g)$

3. $Ca(s)$ + $O_2(g)$ → $CaO(s)$

4. $P(s)$ + $Cl_2(g)$ → $PCl_3(l)$

5. $Fe(OH)_3$ → $Fe_2O_3(s)$ + $H_2O(l)$

6. $N_2(g)$ + $H_2(g)$ → $NH_3(g)$

7. $Mg(s)$ + $HCl(aq)$ → $MgCl_2(aq)$ + $H_2(g)$

8. $Pb(NO_3)_2(s)$ → $PbO(s)$ + $NO_2(g)$ + $O_2(g)$

Activity C1e.6 Reacting masses

Learning objectives

- To provide practise with reacting mass calculations
- To introduce the concept of molarity as a means of denoting concentration
- To introduce the concept of percentage yield

Learning outcomes

- Be able to use chemical equations to calculate reacting masses and the amounts of products formed
- Understand how to carry out mole calculations using volumes and molar concentrations of solutions
- Be able to calculate percentage yields in reactions by comparing actual amounts produced with those predicted by the chemical equation
- Be able to make observations and measurements, analyse and interpret data and evaluate data and methods

Resources

Student Book pages 61–64

Worksheet C1e.6a Obtaining zinc oxide from zinc carbonate, Worksheet C1e.6b Chemical calculations

Files on CD–ROM: C1e.6a_worksheet, C1e.6b_worksheet, C1e.6a_tech_notes

Resources for a class practical (see below)

Approach

1. Decomposition of zinc carbonate

The first part of the activity involves the decomposition of zinc carbonate. It reinforces the idea that you can compare experimental results with theoretical results derived from the chemical equation.

Make this into a competition by dividing students into groups for a class practical. Assign each group a particular starting mass of zinc carbonate. Worksheet C1e.6a gives details of the experiment.

The number of spatula measures covered by the groups should range from 1 to 4. Two groups might have the same starting quantity. The worksheet gives a context for introducing percentage yield.

2. Calculations

The second part of the activity offers additional work on reacting masses and introduces molarity. The Student Book pages 62–64 provide worked examples and Worksheet C1e.6b has related questions.

Decide how much time to spend on this. Students may need to revisit reacting mass calculations several times before they become confident. There are opportunities to do this in the rest of the course.

Technician's notes

You will need the following resources for a class practical.

Be sure to check the latest safety notes on these resources before proceeding.

dry test tubes
test tube holders
mineral wool
access to a balance accurate to 2 decimal places

spatula
zinc carbonate
eye protection
gloves or tongs

Answers

Page 63

1. 1 mole of $CaCO_3$ (40 + 12 + 16 + 16 +16 = 100 g) produces I mole of CaO (40 + 16 = 56 g)

   So 50/100 = 0.5 mole will produce 0.5 mole of CaO = 28 g

2. a) 2 moles

   b) 0.01 moles (100/1000 × 0.1)

   c) 0.25 moles (500/1000 × 0.5)

C1e.6a Obtaining zinc oxide from zinc carbonate

When zinc carbonate is heated, zinc oxide and carbon dioxide are produced. Your task is to find out the relationship between the mass of zinc carbonate and the mass of zinc oxide that can be obtained from it.

zinc carbonate → zinc oxide + carbon dioxide

$ZnCO_3(s) \rightarrow ZnO(s) + CO_2(g)$

(A_r: C = 12; O = 16; Zn = 65)

Apparatus

dry test tube

test tube holder

mineral wool

access to a balance (accurate to 2 decimal places)

eye protection

gloves or tongs

SAFETY INFORMATION

Wear eye protection.

Wear gloves or use tongs to handle the mineral wool.

Method

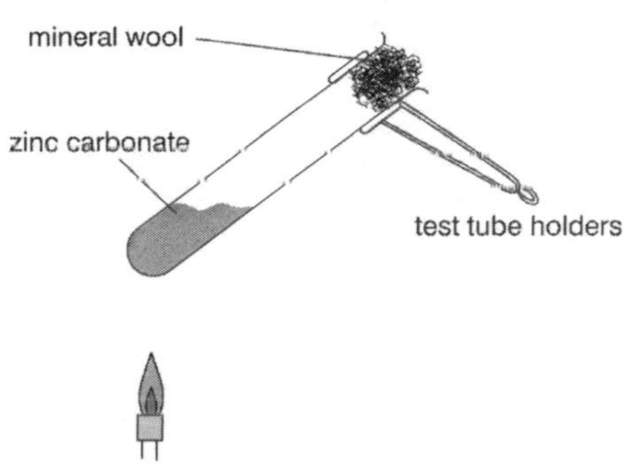

1. Put on eye protection and gloves (or have tongs to hand).
2. Weigh a clean, dry test tube and a plug of mineral wool. Use tongs or gloves to handle the mineral wool.
3. Add a number of spatula measures of zinc carbonate to the tube (your teacher will tell you how many). Reweigh the tube containing the zinc carbonate and the mineral wool.
4. Heat the zinc carbonate gently at first and then more strongly until it has all turned yellow. Then allow the tube to cool before reweighing the tube, zinc oxide and mineral wool.

Make observations and measurements

Mass of test tube + mineral wool + zinc carbonate = ___________ g

Mass of test tube + mineral wool + zinc oxide = ___________ g

Mass of test tube + mineral wool = ___________ g

Mass of zinc carbonate = ___________ g

Mass of zinc oxide = ___________ g

Class results

Group	Mass of zinc carbonate (g)	Mass of zinc oxide (g)
1		
2		
3		
4		
5		
6		
7		
8		

Analyse and interpret data

1. Draw a graph of mass of zinc carbonate (*y*-axis) against mass of zinc oxide (*x*-axis). Include the origin (0, 0) on the graph. Add all the class results and draw a line of best fit through the origin and the class results.

2. Use the equation to calculate the mass of zinc oxide that should be produced from 2.5 g of zinc carbonate (the theoretical yield). Add your result to the graph and draw another line from the origin to this point.

Evaluate data and methods

1. Why was the mineral wool used?

2. How do the class results compare to the line showing the theoretical yield?

3. Suggest reasons for any differences between the line of best fit for the class results and the theoretical yield line.

C1e.6b Chemical calculations

1. What is the concentration of the following solutions?
   a) A solution containing 2 moles of solute in 2 dm^3 of solution
   b) A solution containing 2 moles of solute in 4 dm^3 of solution
   c) A solution containing 1 mole of solute in 10 dm^3 of solution
   d) A solution containing 0.1 moles of solute in 500 cm^3 of solution
   e) A solution containing 0.5 moles of solute in 250 cm^3 of solution

2. How many moles of solute are there in the following solutions?
   a) 1 dm^3 of 2M solution
   b) 3 dm^3 of 0.5M solution
   c) 0.5 dm^3 of 2M solution
   d) 100 cm^3 of 4M solution
   e) 250 cm^3 of 1M solution

3. What is the percentage yield obtained in the following experiments?
   a) 1.5 g of calcium carbonate was obtained when the theoretical amount was 2.0 g.
   b) 14 tonnes of ammonia was obtained when the theoretical amount was 34 tonnes.
   c) 1.2 tonnes of sodium hydroxide was obtained when the theoretical amount was 1.5 tonnes.
   d) 2 moles of carbon dioxide was obtained when the theoretical amount was 2.5 moles.

4. Calcium carbonate reacts with dilute hydrochloric acid to produce carbon dioxide.

   $CaCO_3(s) + 2HCl(aq) \rightarrow CaCl_2(aq) + CO_2(g) + H_2O(l)$

   What is the maximum theoretical mass of carbon dioxide that could be made from 10 g of calcium carbonate? Assume the hydrochloric acid is in excess. (Relative atomic masses H = 1; C = 12; O = 16; Ca = 40)

Activity C1e.7 Consolidation and summary

Learning objectives

- To review the learning points of the topic
- To test understanding through answering questions

Learning outcomes

- Be familiar with the knowledge and understanding summarised in the End of Topic Checklist
- Be able to apply this knowledge and understanding by answering the End of Topic Questions

Resources

Student Book pages 65–67

Approach

In planning your approach, you should balance time for consolidation against opportunities for reinforcing the concepts later in the course.

Students' responses to the End of Topic Questions should indicate what further work is needed. Almost all topics will revisit writing and balancing equations, so these should not be a problem.

You should plan carefully for revising the ideas of molarity and reacting masses.

Answers

End of Topic Questions mark scheme

Question	Correct answer	Marks
1	<u>Cu</u> <u>O</u> 0.64 g 0.16 g 0.64/64 0.16/16 0.01 0.01 The simplest formula is CuO	 1 mark 1 mark
2 a)	<u>Na</u> <u>Br</u> 2.3 g 8.0 g 2.3/23 8.0/80 0.1 0.1 The simplest formula is NaBr	 1 mark 1 mark
2 b)	<u>C</u> <u>O</u> 0.6 g 1.6 g 0.6/12 1.6/16 0.05 0.10 The simplest formula is CO_2	 1 mark 1 mark

Question	Correct answer	Marks
2 c)	<u>Fe</u> <u>Cl</u> 11.12 g 21.08 g 11.12/56 21.08/35.5 0.20 0.59 The simplest formula is $FeCl_3$	 1 mark 1 mark
3	<u>Ti</u> <u>Cl</u> 25 g 75 g 25/48 75/35.5 0.52 2.11 The simplest formula is $TiCl_4$	 1 mark 1 mark 1 mark
4	Empirical formula mass of ethene = 12 + 1+1 = 14 g Molecular formula = CH_2 x 28/14 = C_2H_4	1 mark 1 mark
5 a)	<u>C</u> <u>H</u> 92.3 g 7.7 g 92.3/12 7.7/1 7.7 7.7 The empirical formula is CH	 1 mark 1 mark
5 b)	Empirical formula mass = 12 + 1 = 13 g As molecular formula mass is 26 g, the molecular formula is C_2H_2.	1 mark 1 mark
6 a)	$C(s) + O_2(g) \rightarrow CO_2(g)$	1 mark
6 b)	$4Fe(s) + 3O_2(g) \rightarrow 2Fe_2O_3(s)$	1 mark
6 c)	$2Fe_2O_3(s) + 3C \rightarrow 4Fe(s) + 3CO_2(s)$	1 mark
6 d)	$CaCO_3(s) + 2HCl(aq) \rightarrow CaCl_2(aq) + CO_2(g) + H_2O(l)$	1 mark
7	$2Na(s) + 2H_2O(l) \rightarrow 2NaOH(aq) + H_2(g)$ 2 moles 2 moles 46 g 80 g 46/20 = 2.3 g 80/20 = <u>4.0 g</u>	 1 mark 1 mark 1 mark
8	$BaCl_2(aq) + Na_2SO_4(aq) \rightarrow BaSO_4(s) + 2NaCl(aq)$ 1 mole 1 mole 5000 cm^3 0.2M 233 g 5000/100 = 50 cm^3 0.2M 233/100 = <u>2.33 g</u>	 1 mark 1 mark 1 mark
9	2.1/2.8 × 100 75%	1 mark 1 mark
10 a)	1/24 = 0.042 moles	1 mark
10 b)	100/1000 = 0.1 moles	1 mark

Question	Correct answer	Marks
10 c)	1 mole of Mg reacts with 2 moles of HCl (from equation)	1 mark
	Therefore 0.042 moles Mg reacts with 0.084 moles of HCl	1 mark
	0.084 moles is less than 0.1 mole, so HCl is in excess	1 mark
10 d)	1 mole of Mg produces 1 mole of $MgCl_2$	1 mark
	24g of Mg produces 95g of $MgCl_2$	1 mark
	1g of Mg produces 3.96g of $MgCl_2$	1 mark
10 e)	1 mole of Mg produces 1 mole of H_2	1 mark
	24g of Mg produces 2g of H_2	1 mark
	1g of Mg produces 0.083g of H_2	1 mark
10 f)	2g H_2 of occupies 24000 cm^3	1 mark
	0.083g of H_2 occupies 1000 cm^3	1 mark
10 g)	3.1/3.96 × 100	1 mark
	78.3%	1 mark
	Total:	44 marks

C1f Ionic compounds

Introduction

This topic builds on previous ideas about atomic structure. Students must understand the nature of ionic bonding and the properties of ionic compounds in order to study a number of topics later in the course. This topic has two main aspects: the nature of ionic bonding and the typical properties of ionic compounds.

Links to other topics

Sections	Essential background knowledge	Useful links
1 Principles of chemistry	1c Atomic structure	1g Covalent substances 1i Electrolysis
2 Chemistry of the elements		2b Group 1 elements 2c Group 7 elements 2g Tests for ions and gases
4 Physical chemistry		4a Acids, alkalis and salts
5 Chemistry in society		5a Extraction and uses of metals

Topic overview

C1f.1	**Orientation** This activity recaps previous work on atomic structure, with emphasis on electron arrangements. Considering the properties of ionic compounds provides an insight into the nature of the bonding in these compounds.
C1f.2	**The formation of ions** This activity involves modelling the process of ionic bonding and then providing plenty of practice producing dot and cross diagrams. Reference to the relationship between ion charges and the Periodic Table will provide a link to future work in that area.
C1f.3	**Properties of ionic compounds** This activity focuses on the characteristic properties of ionic compounds and contrasts them with other compounds (essentially covalent compounds). It also introduces the idea of a giant ionic lattice structure as a representation of how ions are arranged in an ionic compound.
C1f.4	**Consolidation and summary** This activity provides a quick recap of the ideas encountered and an opportunity to answer the Review Questions in the Student Book.

Activity C1f.1 Orientation

Learning objectives

- To revise ideas on atomic structure
- To map out the main purposes of the topic

Learning outcomes

- Know the structure of an atom by interpreting atomic and mass numbers
- Know the arrangement of electrons in shells for the first 20 elements

Common misconceptions

All compounds are made up of molecules.

Resources

Student Book page 68

Approach

Start by revising the ideas on atomic structure, emphasising the electronic structures of atoms and the arrangement of electrons in shells around the nucleus.

Ask students to research the melting points of some common substances, including a range of ionic compounds and covalent compounds. Ask them to see if there is a pattern in the compounds that have melting points in the higher ranges. They should identify the inclusion of a metal as a common feature.

Explain that these properties must be a result of how metals and non-metals combine together – a process that produces particles known as ions.

Activity C1f.2 The formation of ions

Learning objectives

- To illustrate how metal atoms lose electrons and non-metal atoms gain electrons
- To draw dot and cross diagrams to illustrate how atoms combine together

Learning outcomes

- Be able to describe the formation of ions by the gain or loss of electrons
- Understand that oxidation is the loss of electrons and reduction is the gain of electrons
- Know the charges of a range of common ions
- Be able to deduce the charge on an ion from the electronic structure of the atom from which it is formed
- Be able to explain the formation of ionic compounds of metals from Groups 1, 2 and 3 and non-metals from Groups 5, 6 and 7 using dot and cross diagrams

Common misconceptions

All compounds are made up of molecules.

Resources

Student Book pages 68–71

Worksheet C1f.2a Ionic bonding: dot and cross diagrams

Files on CD–ROM: C1f.2a_worksheet

Approach

Revise the idea of atom diagrams. Use the examples on page 69 of the Student Book to introduce dot and cross diagrams.

Emphasise the importance of electron transfer from metal to non-metal. In the examples that students are likely to meet, both the metal and non-metal acquire a full outer shell of electrons (that is 2 or 8). Take time to explain that the resulting particles are no longer atoms (no longer neutral) and are known as ions.

Emphasise that the ion charges in the chemical formula of an ionic compound cancel each other out. Use the questions on page 71 of the Student Book and Worksheet C1f.2a to allow for practice producing dot and cross diagrams.

Ask students if they can see a pattern between the charge on the resulting ions and their position in the Periodic Table.

Answers

Page 71

1.

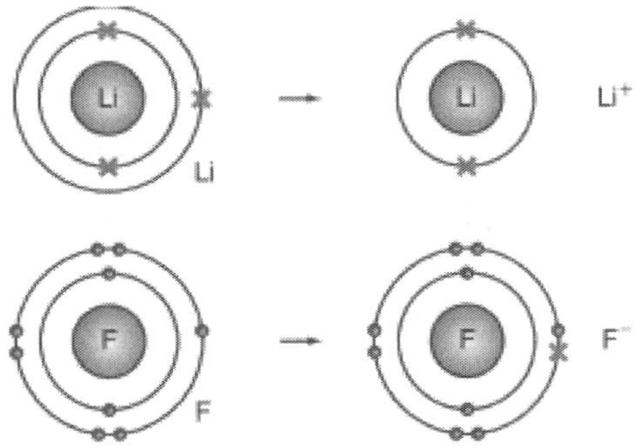

2.

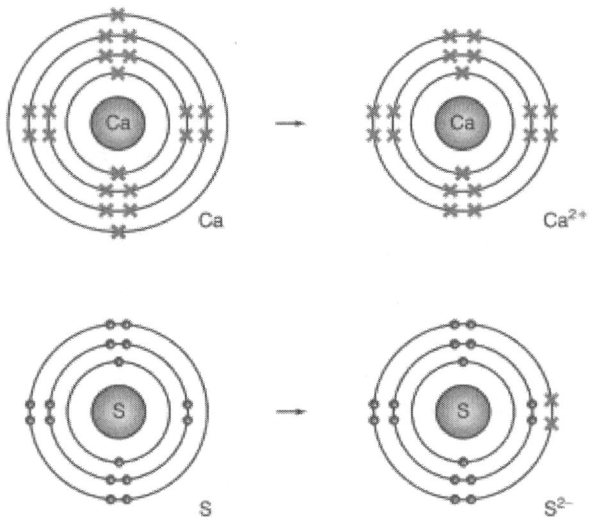

3. No. Both phosphorus and oxygen are non-metals. (A metal is needed to form an ionic bond.)

C1f.2a Ionic bonding: dot and cross diagrams

Ionic bonding is the type of bonding that occurs between metals and non-metals. Metal atoms lose electrons from their outer electron shells and form positive ions; non-metal atoms gain electrons to fill their outer electron shells and form negative ions.

Remember: The number of positive and negative charges must balance.

Draw dot and cross diagrams to show the ionic bonding in the following compounds. Write the formulae of the ions formed in each case.

Potassium oxide (Atomic numbers O = 8; K = 19)
Calcium chloride (Atomic numbers Cl = 17; Ca = 20)
Lithium sulfide (Atomic numbers Li = 3; S = 16)
Aluminium oxide (Atomic numbers O = 8; Al = 13)

Activity C1f.3 Properties of ionic compounds

Learning objectives

- To illustrate the characteristic properties of ionic compounds.
- To explain the effect of the size of ion charges on melting and boiling points

Learning outcomes

- Know that in the solid state the ions of an ionic compound are arranged in a giant ionic 3-dimensional (3D) lattice structure held together by the attractions between oppositely charged ions
- Understand that ionic bonding is a strong electrostatic attraction between oppositely charged ions
- Understand that ionic compounds have high melting and boiling points because of strong electrostatic forces between oppositely charged ions
- Understand how the size of the charges in an ionic compound will determine the relative size of its melting and boiling points
- Be able to draw a simple diagram to show the arrangement of sodium and chloride ions in a crystal of sodium chloride

Resources

Student Book pages 71–73

Approach

If possible, use models to illustrate the 3D structure of ionic compounds such as sodium chloride. Explain that the attractions between positive and negative ions are called electrostatic attractions and that the multiple attractions in all directions produce a very strong structure that is not easily broken down: for example, the melting point of sodium chloride is 801 °C.

Use Table 1.13 on page 72 of the Student Book to summarise the properties of sodium chloride. Mention that these properties apply to all ionic compounds.

Ask students how the properties of magnesium oxide would compare with those of sodium chloride. Explain the higher melting point in terms of a stronger electrostatic attraction between doubly charged ions. You can further exemplify the concept with reference to aluminium oxide.

Answers

Page 73

1. The ions are held together strongly in a giant lattice structure. The ions can vibrate but cannot move around.

2. Sodium chloride is made up of singly charged ions, Na^+ and Cl^-, whereas the magnesium ion in magnesium oxide has a double charge, Mg^{2+}. The higher the charge on the positive ion, the stronger the attractive forces between the positive ion and the negative ion.

Activity C1f.4 Consolidation and summary

Learning objectives

- To review the learning points of the topic
- To test understanding through answering questions

Learning outcomes

- Be familiar with the knowledge and understanding summarised in the End of Topic Checklist
- Be able to apply this knowledge and understanding by answering the End of Topic Questions

Resources

Student Book pages 74–75

Approach

This is a short topic and the ideas covered should require little more than gentle prompting. The End of Topic Checklist in the Student Book can be used to recap the learning outcomes and then the End of Topic Questions can be tackled.

If you have chosen to teach covalent substances as the next topic, then the link can be emphasised.

Answers

End of Topic Questions mark scheme

Question	Correct answer	Marks
1 a)	Not ionic.	1 mark
1 b)	Not ionic.	1 mark
1 c)	Iconic	1 mark
1 d)	Not ionic	1 mark
1 e)	Ionic	1 mark
2 a)	K^+	1 mark
2 b)	Al^{3+}	1 mark
2 c)	S^{2-}	1 mark
2 d)	F^-	1 mark
3	Completed table as shown below	3 marks

Atom	Electronic arrangement of the atom	Electronic arrangement of the ion	Charge on the ion
X	2, 6	2, 8	2–
Y	2, 8, 8, 2	2, 8, 8	2+
Z	2, 1	2	1+ (or +)

4 a)	Dot and cross diagrams should show the following: 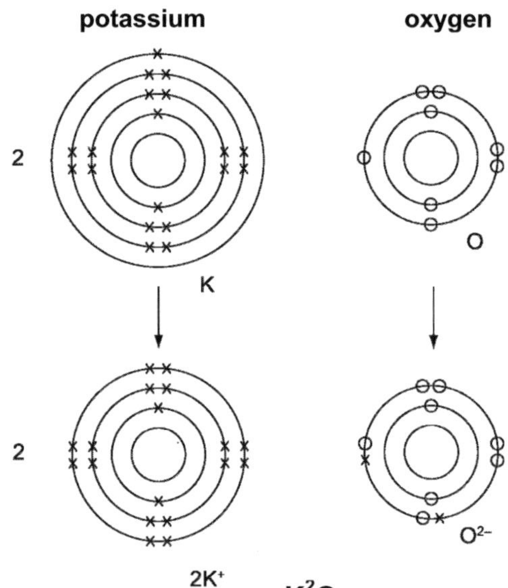 K^2O	2 marks
4 b)	Dot and cross diagrams should show the following: 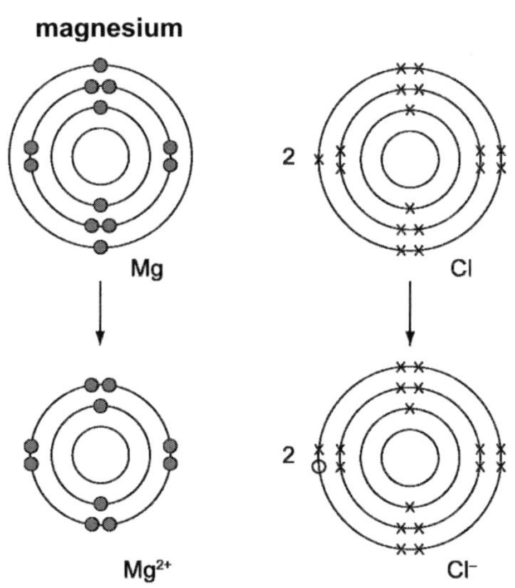 $MgCl_2$	2 marks
5 a)	There are strong electrostatic attractive forces between the ions. A lot of energy needs to be applied to break these attractions to form a liquid.	1 mark 1 mark
5 b)	The ions are able to carry the electric current. The ions need to be free to move, therefore the potassium chloride needs either to be in a molten state or in solution.	1 mark 1 mark
6	Magnesium forms a 2+ ion, oxygen a 2- ion and chlorine a 1- ion. The electrostatic attraction between 2+ and 2- ions is greater than between 2+ and 1- ions resulting in a higher melting point and boiling point.	1 mark 1 mark
	Total:	22 marks

C1g Covalent substances

Introduction

This topic builds on previous discussion about atomic structure. Students must understand the nature of covalent bonding and the properties of covalent compounds in order to study a number of topics later in the course. The topic has three main aspects: the nature of covalent bonding; the typical properties of covalent compounds and the covalent bonding in diamond and graphite.

Links to other topics

Sections	**Essential background knowledge**	**Useful links**
1 Principles of chemistry	C1c Atomic Structure	1f Ionic compounds
3 Organic chemistry		3b Alkanes 3c Alkenes 3d Ethanol
4 Physical chemistry		4b Energetics
5 Chemistry in society		5b Crude oil 5c Synthetic polymers

Topic overview

C1g.1	**Orientation** This activity builds on earlier work on atomic structure and ionic bonding. Covalent bonding is contrasted with that of ionic bonding and the differences in typical properties of covalent and ionic compounds are highlighted.
C1g.2	**The formation of covalent molecules** This activity involves modelling the process of covalent bonding and then providing plenty of practice producing dot and cross diagrams. Reference to the relationship between the number of covalent bonds an atom can form and the Periodic Table will provide a link to future work in that area.
C1g.3	**Properties of covalent compounds** This activity focuses on the characteristic properties of covalent compounds and contrasts them with other compounds (essentially ionic compounds). It also introduces the idea of a relatively weak force existing between molecules in simple molecular substances.
C1g.4	**Diamond and graphite** This activity looks at the nature of the covalent bonding and giant structures in diamond and graphite. Differences in physical properties are explained in terms of their differing structures.
C1g.5	**Consolidation and summary** This activity provides an opportunity for a quick recap the ideas encountered in the topic as well as time for students to answer the End of Topic Questions in the Student Book.

Activity C1g.1 Orientation

Learning objectives

- To revise ideas on atomic structure and ionic bonding
- To map out the main purposes of the topic

Learning outcomes

- Know the structure of an atom from its atomic and mass numbers
- Know the arrangement of electrons in shells for the first 20 elements

Resources

Student Book page 76

Approach

If you choose to teach this topic soon after Topic 1f *Ionic compounds*, you will be able to revise atomic structure very quickly. Refer to the exercise that students did on melting points, and the differences between compounds made up from non-metals and those that contain a metal.

Explain that these differences can be explained in terms of the relative strength of the attractive forces between the ions (in ionic compounds) and between the molecules in covalent compounds. Tell students that it is important to distinguish between the attractive forces within molecules (covalent bonds) and the attractive forces between molecules.

Activity C1g.2 The formation of covalent compounds

Learning objectives

- To illustrate how non-metals share electrons
- To draw dot and cross diagrams to show how non-metal atoms combine together

Learning outcomes

- Know that a covalent bond is formed when electrons are shared between the atoms of two non-metals
- Understand that covalent bonding is a result of a strong attraction between the bonding pair of electrons and the nuclei of the atoms involved in the bond
- Be able to explain the type of covalent bonding using dot and cross diagrams for the following: hydrogen, chlorine, hydrogen chloride, water, methane, ammonia, ethane, oxygen, nitrogen, carbon dioxide and ethene

Resources

Student Book pages 76–79

Worksheet C1g.2a Covalent bonding: dot and cross diagrams

Files on CD–ROM: C1g.2a_worksheet

Approach

Use the examples in the Student Book to illustrate how non-metals, which do not like to give up electrons, share electrons to achieve the electronic configuration of the nearest noble gas (full outer electron shells). Emphasise the importance of showing shared electrons in shared pairs in single and multiple covalent bonds.

It is also a good idea to show non-bonding electrons as electron pairs. This will help students to avoid mistakes in showing the correct number.

Introduce the structure of covalent molecules represented by a displayed formula. At this level it is acceptable to draw methane with a square planar displayed formula, as opposed to a pyramidal one.

Use the questions on page 79 of the Student Book and Worksheet C1g.2a to give students practice at producing dot and cross diagrams.

Ask students if they notice a relationship between the number of covalent bonds an atom can form and its position in the Periodic Table.

Answers

Page 79

1.

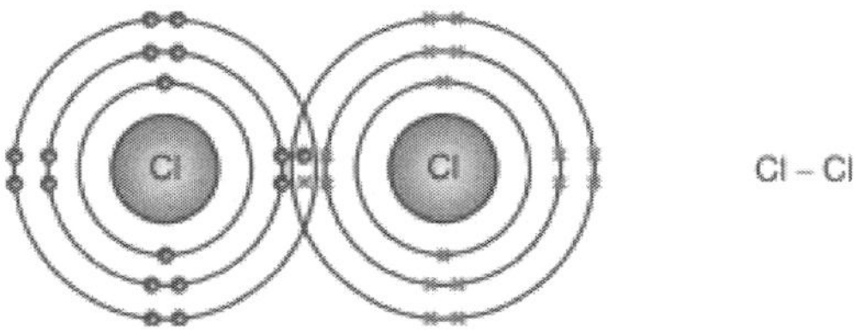

2.

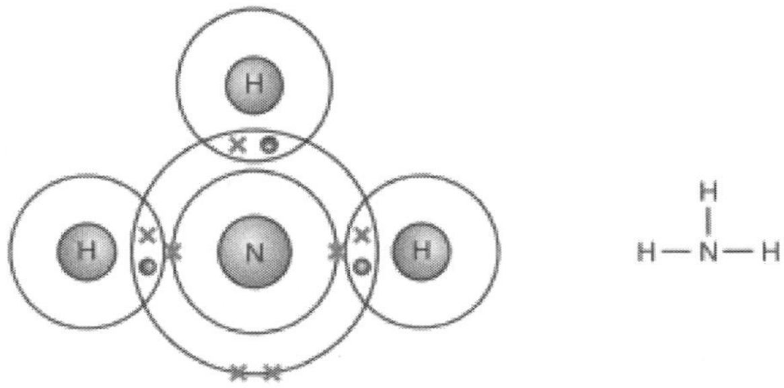

3.

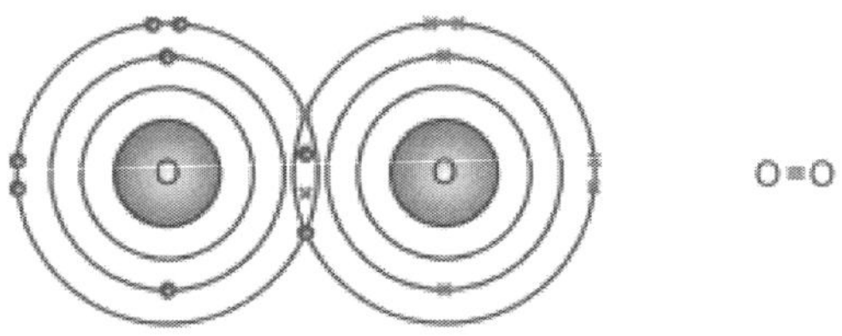

4.

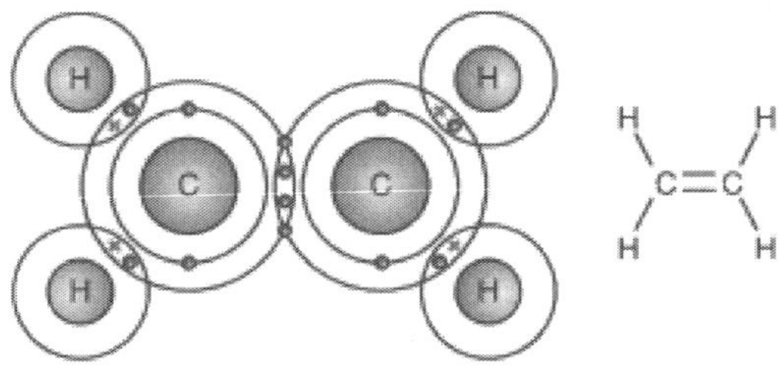

5.

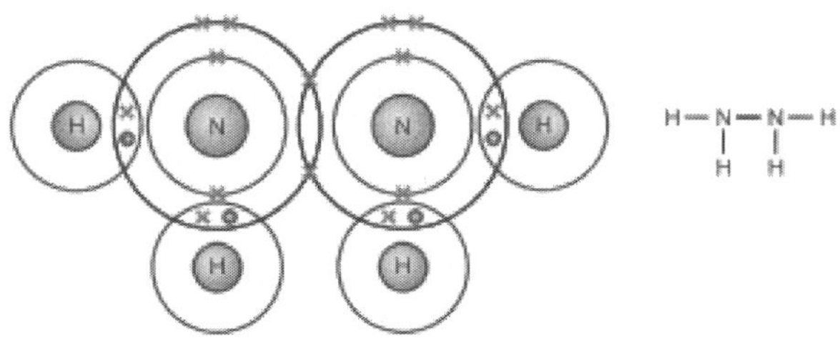

C1g.2a Covalent bonding: dot and cross diagrams

Covalent bonding is the type of bonding that occurs between non-metals. The atoms share electrons from their outer electron shells and form molecules. In most cases you will encounter, the atoms achieve the electron configuration of the nearest noble gas (a full outer shell of electrons) by sharing electrons with one or more other atoms.

Draw dot and cross diagrams to show the covalent bonding in the following compounds. Write the displayed formula of the molecule formed in each case.

Hydrogen sulfide (Atomic numbers H = 1; S = 16)
Chlorine oxide (Atomic numbers O = 8; Cl = 17)
Phosphorus chloride (Atomic numbers P = 15; Cl = 17)
Propane (Atomic numbers H = 1; C = 6; propane contains hydrogen atoms and three carbon atoms)

Activity C1g.3 Properties of covalent compounds

Learning objectives

- To illustrate the characteristic properties of covalent compounds
- To provide an explanation for these properties in terms of the strength of forces between the molecules

Learning outcomes

- Know that simple molecular substances are gases, liquids or solids with low melting points
- Be able to explain the properties of simple molecular substances in terms of the relatively weak forces between the molecules

Common misconceptions

Some students think that a covalent bond is weaker than an ionic bond and that covalent bonds are weak bonds.

Resources

Student Book page 80

Approach

Decide whether to introduce the terms *intramolecular* and *intermolecular* – alternatively you could use *within molecules* and *between molecules*. Many students are confused by the difference, so it is worth spending a little time explaining this.

Use the section in the Student Book on page 80 to identify the characteristic properties of simple covalent compounds. Emphasise the lack of electrical conductivity due to the absence of ions.

Contrast the properties of these simple molecular structures with those characteristic of ionic compounds.

Activity C1g.4 Diamond and graphite

Learning objectives

- To illustrate the nature of the covalent bonding in diamond and graphite and to consider their differing structures
- To use the structures to explain the characteristic properties of covalent molecules

Learning outcomes

- Know that giant covalent structures contain a lattice of strong covalent bonds
- Be able to draw simple diagrams showing the arrangement of carbon atoms in diamond and graphite
- Know the uses of diamond (as a cutting edge) and graphite (as a lubricant) and can explain these uses in terms of the structures of the two substances

Common misconceptions

Some students think that graphite is soft.

Resources

Student Book pages 80–82

Approach

The structures and properties of diamond and graphite (such as their melting points) make the point very convincingly that covalent bonds are strong bonds.

If possible, use models of the two structures to identify similarities and differences. Refer to the diagrams and explanations in the Student Book on page 81 and the Science in Context feature on diamonds.

Answers

Page 80

1. The intermolecular forces of attraction between the molecules are weak.

2. No. There are no ions or delocalised electrons present.

Page 82

1. Each carbon atom is strongly covalently bonded to four other atoms forming a very strong giant lattice structure. A very high temperature is required to break down the structure.

2. In graphite each carbon atom is strongly covalently bonded to 3 other carbon atoms. The remaining outer shell carbon electron is delocalised and so can move along the layers formed by the covalently bonded carbon atoms.

Activity C1g.5 Consolidation and summary

Learning objectives

- To review the learning points of the topic
- To test understanding through answering questions

Learning outcomes

- Be familiar with the knowledge and understanding summarised in the End of Topic Checklist
- Be able to apply this knowledge and understanding by answering End of Topic Questions

Resources

Student Book pages 83–84

Approach

This is a short topic and so the ideas covered should require little more than gentle prompting. The End of Topic Checklist can be used to recap the learning outcomes and then the End of Topic Questions can be tackled.

Ask students to produce a table summarising the differences between the bonding and properties of ionic and covalent compounds.

Answers

End of Topic Questions mark scheme

Question	Correct answer	Marks
1 a)	Dot and cross diagram of hydrogen fluoride showing: 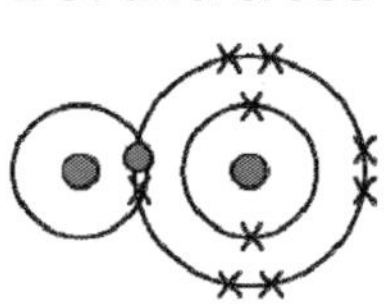 H – F (1 mark for the correct electron arrangements, 1 mark for correct sharing of electrons)	2 marks
1 b)	Dot and cross diagram of carbon disulfide showing: 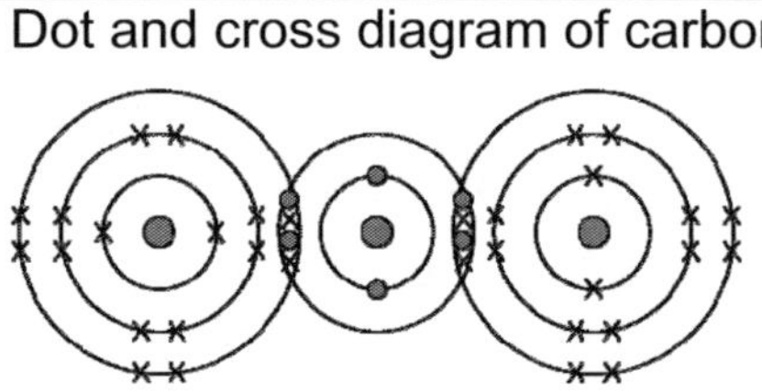 S = C = S (1 mark for the correct electron arrangements, 1 mark for correct sharing of electrons)	2 marks
1 c)	Dot and cross diagram of ethanol. It should be the same as the dot and cross diagram for ethane:	2 marks

Question	Correct answer	Marks
	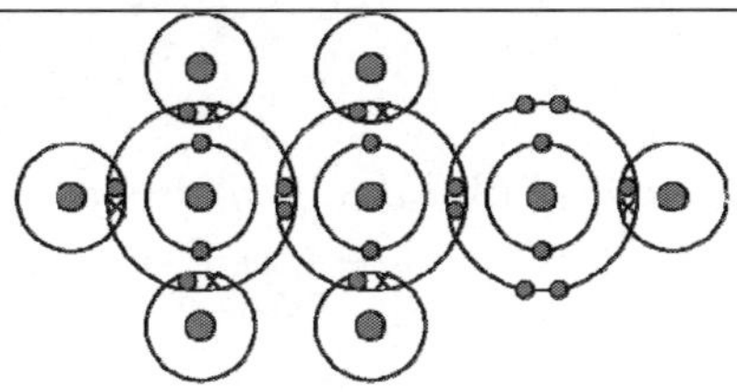 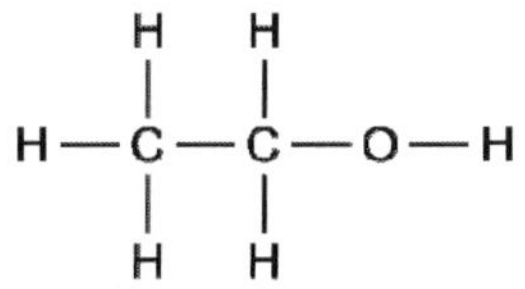 (1 mark for the correct electron arrangements, 1 mark for correct sharing of electrons)	
2	The forces of attraction between the candle wax molecules are (relatively) weak. The energy needed to break these forces is (relatively) low.	1 mark 1 mark
3	The attractive forces between the particles (atoms or molecules) in a gas are (relatively) weak. In an ionic compound the attractive forces between the particles (ions) are (relatively) strong.	1 mark 1 mark
4	The attractive forces between the methane molecules are (relatively) weak. A (relatively) small amount of energy at the melting point is needed to break these attractive forces in solid methane.	1 mark 1 mark
5 a)	X is likely to be a liquid or a gas. In a simple molecular structure the attractive forces between the particles (atoms or molecules) are (relatively) weak, as in a liquid or gas.	1 mark 1 mark
5 b)	X is likely to have a (relatively) low boiling point. The attractive forces between the particles in liquid X are (relatively) weak and so not much energy is needed at the boiling point to break them.	1 mark 1 mark
6 a)	The layers of carbon atoms in graphite are held together by strong covalent bonds. As the covalent bonds in the layers are not easily broken, they will add strength to the tennis racquet.	1 mark 1 mark
6 b)	Between the covalently bonded layers in the graphite are delocalised electrons. These are able to move and carry an electric current.	1 mark 1 mark
7 a)	Diamond is a giant (atomic) structure with each carbon atom covalently bonded to 4 other carbon atoms. The lattice of strong covalent bonds gives diamond its hardness.	1 mark 1 mark
7 b)	Diamond is used in drills/cutting tools.	1 mark
	Total:	23 marks

C1h Metallic crystals

Introduction

This is a very short topic. Some, if not all, of the activities could be combined into a single lesson. You can link this topic to atomic structure and ionic bonding, but there is a one new topic to cover: the model illustrating the structure of metals.

In the teaching sequence the topic is likely to follow on from work on ionic and covalent bonding, so finishing the work on structure and bonding. Later in the course you can reinforce students' understanding of metallic bonding.

Links to other topics

Sections	Essential background knowledge	Useful links
1 Principles of chemistry	1c Atomic structure 1f Ionic compounds	
2 Chemistry of the elements		2b Group 1 elements 2f Reactivity series
5 Chemistry in society		5a Extraction and uses of metals

Topic overview

C1h.1	**Orientation** This activity builds on the work done on atomic structure and ionic bonding and reflects on the properties of metals and the structural features that can explain the properties.
C1h.2	**Metallic bonding** This is the main activity of the topic. The model for the structure of a metal is introduced and then used to explain the characteristic properties of metals. Alloys are also included as a means of optimising the properties of a metal.
C1h.3	**Consolidation and summary** This activity allows for a very quick recap on the ideas encountered and gives time for the students to answer the End of Topic Questions in the Student Book.

Activity C1h.1 Orientation

Learning objectives

- To summarise knowledge and understanding of atomic structure and ionic bonding
- To map out the main purposes of the topic

Learning outcomes

- Know that an atom is made up of a positively charged nucleus containing protons and neutrons (except hydrogen) with electrons in shells around the nucleus
- Know that metal atoms lose electrons when they form ionic compounds

Resources

Student Book page 85

Approach

If you taught Topic 1f *Ionic compounds* some time ago, this topic provides a good opportunity to revise the key features of ionic bonding.

Ask students to work in small groups. Ask them to write down quickly five things they know about ionic bonding and ionic compounds.

Collect together their responses and focus on what they have said about the role of metal atoms in ionic bonding, such as: they lose electrons. Explain that this easy release of electrons is the key to the properties of metals.

Tell students that the topic will also focus on alloys. Ask them what an alloy is.

Activity C1h.2 Metallic bonding

Learning objectives

- To introduce a model illustrating the structure of a metal
- To use the model to explain the typical physical properties of metals
- To undertake some research of alloys

Learning outcomes

- Be able to describe a metal as a giant structure of positive ions surrounded by a 'sea' of delocalised electrons
- Be able to explain the electrical conductivity of metals in terms of the sea of delocalised electrons
- Be able to explain the malleability of a metal in terms of its structure and bonding
- Know that an alloy is a mixture of a metal and one or more other elements

Resources

Student Book pages 85–87

Approach

Using page 86 of the Student Book, introduce the model that describes the structure of a metal. Emphasise that the individual atoms release their outermost electrons to form a 'sea' of electrons. Introduce the term 'delocalised' to describe this process.

Electrostatic attraction between the nuclei and the sea of electrons is relatively strong, and this explains the relatively high melting points of most metals. You may wish to refer to the Group 1 metals as an exception to this general rule.

Make sure that students understand the terms malleable and ductile and explain these properties in terms of the model. Ask them to research some examples of alloys. For each alloy, ask them to find out:

- the approximate composition
- its uses
- the benefits of using the alloy rather than the pure metal.

Answers

Page 86

1. A cation is a positive ion.

2. Metals contain delocalised electrons that are not fixed to a particular atom. They can move throughout the structure.

3. The structure is not rigid, the ions can move into different positions when it is bent.

Activity C1h.3 Consolidation and summary

Learning objectives

- To review the learning points of the topic
- To test understanding through answering questions

Learning outcomes

- Be familiar with the knowledge and understanding summarised in the End of Topic Checklist
- Be able to apply this knowledge and understanding by answering the End of Topic Questions

Resources

Student Book pages 88–89

Approach

This is a very short topic and may not need a separate consolidation activity. Nevertheless, students will benefit from reflecting on what they have learned. Answering the End of Topic Questions on page 89 of the Student Book should support this.

Answers

End of Topic Questions mark scheme

Question	Correct answer	Marks
1	A metal structure is made up of metal ions. The ions are surrounded by a 'sea' of delocalised electrons.	1 mark 1 mark
2 a)	Metals contain a 'sea' or 'cloud' of delocalised electrons. The electrons are able to move and carry the electric current.	1 mark 1 mark
2 b)	The structure is held together by the attractive forces between the ions and the electrons. Beating the metal disrupts the arrangement of the ions but does not break the attractive forces between the ions and the electrons.	1 mark 1 mark
3	In the graphite structure the delocalised electrons are between the layers of covalently bonded carbon atoms. The electrons are not delocalised over the whole structure and so can only conduct electricity in one plane. In a metal the electrons are delocalised over the whole metal structure and so can conduct in all planes.	1 mark 1 mark 1 mark
4 a)	C	1 mark
4 b)	A	1 mark
4 c)	D	1 mark
4 d)	B	1 mark
5 a)	An alloy is a mixture of a metal and other elements.	1 mark
5 b)	They modify the property of the pure metal making it more useful. Alloys can be stronger/more hard-wearing than the pure metal (e.g. steel).	1 mark 1 mark
	Total:	16 marks

C1i Electrolysis

Introduction

This topic builds on the understanding of chemical bonding gained from previous work on ionic compounds and covalent substances. It also provides an opportunity to revise the structure of metals and their property as conductors of electricity.

The reactivity series can be used to predict the preferential discharge of ions. The concepts of electrolysis can be reinforced by practical demonstrations. This also gives students some good opportunities to develop their investigative skills, particularly *Make observations and measurements* and *Analyse and interpret data*. The ideas will be reinforced in the Section 5 *Chemistry in society*.

Links to other topics

Sections	Essential background knowledge	Useful links
1 Principles of Chemistry	1f Ionic compounds 1g Covalent substances 1h Metallic crystals	
2 Chemistry of the elements	2f Reactivity series	
5 Chemistry in society		5a Extraction and uses of metals 5d The industrial manufacture of chemicals

Topic overview

C1i.1	**Orientation** This activity revises ideas on chemical bonding, especially on ionic bonding, and then introduces some of the terms associated with electrolysis. Demonstrations highlight the aspects to be covered in the topic.
C1i.2	**Electrolysis of molten electrolytes** This activity introduces the electrolysis of molten electrolytes using lead(II) bromide as an example. The need for the ions in an electrolyte to be able to move is emphasised and ionic half-equations are introduced.
C1i.3	**Electrolysis of aqueous electrolytes** This is a relatively long activity (with four parts) and offers the opportunity for two practical activities for students. The dissociation of water into ions is introduced and used to explain the products formed from the electrolysis of aqueous solutions. Further emphasis is placed on the use of half-equations to summarise the changes that occur at the electrodes. The reactivity series is introduced to explain the preferential discharge of ions at the cathode.
C1i.4	**Quantitative electrolysis** This activity introduces the faraday as a mole of electrons and links the concept to half-equations. A practical activity requires careful quantitative work and will help students to become more familiar with the use of the faraday.
C1i.5	**Consolidation and summary** This activity provides an opportunity for a quick recap of the ideas encountered in the topic.

Activity C1i.1 Orientation

Learning objectives

- To revise ideas on chemical bonding (especially ionic bonding) and metallic bonding
- To introduce the terminology associated with the process of electrolysis
- To map out the main purposes of the topic

Learning outcomes

- Know that the anode is a positive electrode
- Know that the cathode is a negative electrode
- Understand that an electric current is a flow of electrons in a conductor or ions in an electrolyte
- Know that electrolysis is the breaking down of a compound by the passage of electricity (an electric current)
- Know that an electrolyte is a substance that, when molten or dissolved in water, allows an electric current to pass through it

Common misconceptions

Some students think that electrons move within the electrolyte (as opposed to ions).

Resources

Student Book pages 90–92

Files on CD–ROM: C1i.1a_tech_notes

Resources for demonstration (see below)

Approach

Demonstrate the passage of electricity through a solid conductor such as a strip of copper. Contrast it with the fact that a strip of plastic acts as an insulator.

Ask why a metal conducts electricity but plastic does not. Check that students remember that in metals there is a delocalised cloud of electrons that allows electrons to flow through the solid when a potential difference (voltage) is applied.

Then demonstrate the passage of electricity through a solution (such as aqueous sodium chloride) and compare it with the lack of conductivity of solid sodium chloride. Ask students what they can remember about the bonding in sodium chloride. Suggest that the ions in sodium chloride might have something to do with the fact that an electric current can pass through it.

Also ask what will be the difference between the ions in solid sodium chloride and in sodium chloride solution. Use this demonstration to introduce some terms and definitions for: electrode, anode, cathode, electrolyte, electrolysis.

Tell students that the topic will explore further the involvement of ions in electrolysis and why they must be able to move. They should have observed a gas being produced in the electrolysis of sodium chloride solution. Explain that the topic will also be investigating the relationship between the quantity of electricity and the amount of product produced in electrolysis.

Technician's notes

You will need the following resources for a demonstration.

Be sure to check the latest safety notes on these resources before proceeding.

copper foil
strip of plastic
solid sodium chloride
0.5M sodium chloride solution
power supply, bulb, leads, crocodile clips, carbon electrodes

Answers

Page 92

1. The breaking down (decomposition) of a chemical compound by the use of electricity.

2. The positive electrode is the anode.

3. The substance must contain ions and they must be free to move (in molten/liquid state or dissolved in water).

Activity C1i.2 Electrolysis of molten electrolytes

Learning objectives

- To demonstrate the electrolysis of a molten electrolyte, lead(II) bromide
- To emphasise the requirement for ions to be able to move if electrolysis is to take place
- To introduce half-equations

Learning outcomes

- Know that for a substance to act as an electrolyte the ions must be free to move
- Be able to describe the electrolysis of molten substances such as lead(II) bromide
- Be able to write half-equations to represent the reactions that occur at the electrodes

Resources

Student Book pages 92–93

CD–ROM: C1i.2a_tech_notes

Resources for demonstration (see below)

Approach

Revise the conditions for electrolysis discussed in the previous activity. The section on page 91 of the Student Book should help.

Then introduce the electrolysis of lead(II) bromide. You should set this up in a working fume cupboard with apparatus similar to that shown in the Student Book. If possible, it is better to have a lamp in the circuit.

Demonstrate that an electric current cannot pass through the lead(II) bromide as a solid, but once it is molten, electric current does flow. It should be possible for students to see the bromine vapour being formed at the anode and, after cooling, a bead of lead will have formed at the bottom of the crucible. Introduce the half-equations for the reactions that have occurred.

Ask students to write half equations to represent the changes that occur when some other ionic compounds are electrolysed, for example sodium chloride, magnesium chloride and aluminium oxide.

If you need to reinforce the fact that not all molten substance conduct electricity, you could repeat the experiment using candle wax (a covalently bonded compound).

Technician's notes

You will need the following resources for a demonstration.

Be sure to check the latest safety notes on these resources before proceeding.

The apparatus should be set up in a fume cupboard.

carbon electrodes in a holder
electrical leads, crocodile clips, lamp, power pack
tripod, pipe clay triangle, retort stand and clamp
lead(II) bromide
possibly also candle wax

Activity C1i.3 Electrolysis of aqueous electrolytes

Learning objectives

- To allow the students to identify electrolytes from a range of liquids and solutions
- To provide an opportunity to use knowledge of ionic and covalent bonding to make predictions about potential electrolytes

Learning outcomes

- Be able to describe the electrolysis of the following electrolytes and predict the products formed:
  aqueous sodium chloride solution
  aqueous copper(II) sulfate and dilute sulfuric acid
- Be able to write half-equations to represent the reactions that occur at the electrodes

Common misconceptions

Water is an electrolyte (perhaps arising from knowledge of the dangers associated with electrical appliances in the bathroom).

Resources

Student Book pages 93–97

Worksheet C1i.3a Identifying and investigating electrolytes

Worksheet C1i.3b The electrolysis of sodium chloride solution

Files on CD–ROM: C1i.3a_worksheet, C1i.3b_worksheet, C1i.3a_tech_notes, C1i.3b_tech_notes

Resources for class practical (see below)

Approach

1. Investigating electrolytes

Based on the first two activities, students should be able to test a range of liquids and solutions as indicated in Worksheet C1i.3a, making predictions before they undertake each test.

They will become aware that effervescence occurs at one electrode at least with all four electrolytes. They should be able to identify copper as a pink deposit on the cathode.

2. Role of water in electrolysis

Now focus on the role of water in the electrolysis of these solutions – for example, why wasn't sodium formed when sodium chloride was electrolysed?

The slight dissociation of water into ions is covered in the Student Book on pages 93–94. This may be the first time that students have come across the reversible sign and the idea of an equilibrium reaction.

Emphasise that there are too few ions present in water for it to behave as an electrolyte, but that more water molecules dissociate as ions are discharged at electrodes. Take some time to explain the half-equations for the formation of hydrogen, oxygen and chlorine.

3. Electrolysis of sodium chloride solution

Students can now use Worksheet C1i.3b to look in more detail at the electrolysis of sodium chloride solution. This activity introduces the tests for hydrogen and chlorine.

If these have not been covered previously, you will need to provide some guidance.

4. Electrolysis of copper(II) sulfate

Page 95 of the Student Book can be used to support the study of the electrolysis of copper(II) sulfate in the presence of dilute sulfuric acid.

Refer back to the observations the students made for this reaction in Activity C1i.3a. This activity can be completed by focusing on the activity series as a means of identifying which of competing ions will be preferentially discharged.

Developing Investigative Skills on page 97 of the Student Book can be used to consolidate the ideas encountered in the whole activity.

For electrolysis of molten salts; lead bromine, chloride and iodide, you need to wear goggles and use a fume cupboard. These salts are appreciably volatile at melting temperatures. Impure lead(II) bromide may give off some bromine on melting as well as during electrolysis and the pure iodide starts to decompose just above melting point. Consider using anhydrous zinc chloride as a safer alternative.

Technician's notes

You will need the following resources for a class practical.

Be sure to check the latest safety notes on these resources before proceeding.

C1i.3a Identifying and investigating electrolytes

power pack, leads, crocodile clips, lamp, carbon electrodes
100 cm^3 beaker
ethanol (+ container for collection of used ethanol)
dilute hydrochloric acid (0.5M)
dilute sulfuric acid (0.5M)
copper(II) sulfate solution (0.5M)
sodium chloride solution (0.5M)
eye protection

C1i.3b The electrolysis of sodium chloride solution

electrolysis cell with carbon electrodes
power supply, leads, crocodile clips
ignition tubes
sodium chloride solution (0.5M)

Answers

Page 96

1. a) An inert electrode is an unreactive electrode. It will not be changed during electrolysis.

   b) Carbon is commonly used as an inert electrode.

2. a) Lead and chlorine

   b) Magnesium and oxygen

   c) Aluminium and oxygen

3. a) Hydrogen (sodium is above hydrogen in the reactivity series)
   b) Hydrogen (zinc is above hydrogen in the reactivity series)
   c) Silver (silver is below hydrogen in the reactivity series)
4. a) $2O^{2-}(aq) \rightarrow O_2(g) + 4e^-$
   b) The change takes place at the anode.

Developing investigative skills, page 97

1. Platinum is commonly used as a material for inert electrodes.
2. Hydrogen was formed at the cathode.
3. Fill the tubes to overflowing with the electrolyte, slide a small piece of paper over the end of the tube and invert the tube carefully. Carefully put the inverted tube in the dilute sulfuric acid above the electrode (the small piece of paper will float away).
4. Sulfuric acid solution is an electrolyte.
5. The volume of oxygen will be half that of the hydrogen.
6. The ions in dilute sulfuric acid are: H^+; SO_4^{2-} and OH^-.
7. $2H^+(aq) + 2e^- \rightarrow H_2(g)$

C1i.3a Identifying and investigating electrolytes

An electrolyte is a substance that allows an electric current to pass through it when it is molten or dissolved in water. In addition, the current breaks down the electrolyte, producing other chemicals. In this experiment you will test some liquids and solutions to see if they are electrolytes. Your knowledge of chemical bonding should help you to make predictions before you test each liquid/ solution. For each electrolyte you will need to write down what you observe at each electrode as the current is flowing through it.

Apparatus

power pack or 6 V battery

bulb and holder

leads

crocodile clips

carbon electrodes

100 cm^3 beaker

substances to test: dilute hydrochloric acid, dilute sulfuric acid, dilute copper(II) sulfate, dilute sodium chloride solution, ethanol

eye protection

SAFETY INFORMATION
Ethanol is highly flammable.

Method

1. Put on your eye protection.

2. Set up the apparatus as shown in the diagram. A 6 volt DC supply of electricity is suitable. The bulb will indicate if the test liquid or solution is an electrolyte or not.

3. Wipe the electrodes between tests.

4. Rinse and dry the beaker between tests but do NOT tip ethanol into the sink, pour it into the container provided by the teacher.

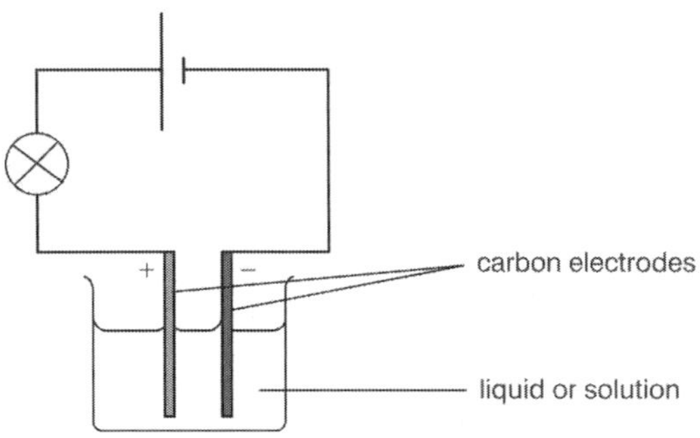

Make observations and measurements

1. Record your results in the table provided

Liquid/solution	Prediction Electrolyte (Y/N)	Electrolyte observations	
		At the anode	At the cathode
Water			
Dilute hydrochloric acid			
Dilute sulfuric acid			
Copper(II) sulfate solution			
Ethanol			
Sodium chloride solution			

Analyse and interpret data

1. Which liquids or solutions were NOT electrolytes?
2. Which electrolyte produced a solid coating on one of the electrode? At which electrode did this coating form?
3. Explain why some of the liquids or solutions did not act as electrolytes whereas others did.

C1i.3b The electrolysis of sodium chloride solution

Sodium chloride is an electrolyte and so it will allow an electric current to pass through it when in solution. It will also be broken down by this electric current. In this experiment you will be able to predict the products of the electrolysis and then test your predictions.

Apparatus

electrolysis cell with carbon electrodes

power supply 6 V DC

electrical leads, crocodile clips

sodium chloride solution

eye protection

SAFETY INFORMATION
Wear eye protection.
Do not run for a long time, otherwise a lot of chlorine gas will be evolved.

Method

1. Put on your eye protection.
2. Using the apparatus shown carefully fill the electrolysis cell about two-third full with sodium chloride solution.
3. Fill the ignition tubes with sodium chloride solution and carefully invert them over the carbon electrodes.
4. Connect the circuit and record your observations.

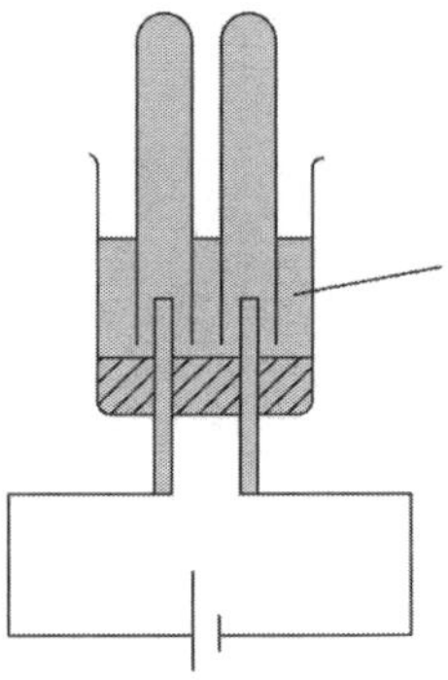

Make observations and measurements

1. Record your observations in the table below:

Observations at the anode	Observations at the cathode

Analyse and interpret data

1. The ion that is discharged at the cathode will have what charge? What do you think this ion is?
2. What do you think is the gas produced at the cathode? What test could you do on the gas to check your prediction? What result would confirm your prediction?
3. The ion that is discharged at the anode will have what charge? What do you think the ion is?
4. What do you think is the gas produced at the anode? What test could you do on the gas to check your prediction? What result would confirm your prediction?
5. Why do you think so little gas collects at the anode?

Activity C1i.4 Quantitative electrolysis

Learning objectives

- To introduce the link between half equations and the mole of electrons (the faraday)

Learning outcomes

- Be able to write half-equations to represent the reactions that occur at electrodes
- Know that a faraday is a mole of electrons
- Understand how to calculate the amounts of products formed in the electrolysis of molten compounds and aqueous solutions from the half equations involved

Resources

Student Book page 98

Worksheet C1i.4a To find out how many faradays of electricity are needed to produce 1 mole of copper

Files on CD-ROM: C1i.4a_worksheet, C1i.4a_tech_notes

Resources for class practical (see below)

Approach

Introduce the faraday as a mole of electrons and relate this to some of the half-equations already encountered in the topic. Use the questions in the Student Book to provide practice for students.

Worksheet C1i.4a provides quantitative work related to the quantity of electricity needed to produce 1 mole of copper.

The calculation of the quantity of electricity from current (I) and time (t) is not a requirement of the course specification, but students should be able to cope with the necessary calculations. It provides a very useful reinforcement of the link between the half-equation and moles of electrons.

Technician's notes

You will need the following resources for a class practical.

Be sure to check the latest safety notes on these resources before proceeding.

power supply, rheostat, ammeter, leads, crocodile clips
balance accurate to 2 decimal places
stopwatch
2 copper electrodes
sandpaper or emery cloth
copper(II) sulfate solution (0.5M)
100 cm^3 beaker

Answers

Page 98

1. 2 faradays of electricity are needed to produce 1 mole of calcium atoms.
2. 2 moles of chlorine gas will be produced by 4 faradays of electricity.
3. 0.5 faradays will produce 0.25 moles of hydrogen gas (H_2) = 0.5 g ($2H^+(aq) + 2e^- \rightarrow H_2(g)$)

C1i.4a To find out how many faradays of electricity are needed to produce 1 mole of copper

When copper(II) sulfate solution is electrolysed, copper is deposited at the cathode. Unfortunately, the copper often falls off the cathode and is difficult to collect and weigh. If the electrolysis is done with copper electrodes, the amount of copper deposited at the cathode is exactly the same as the amount of copper which dissolves at the anode. It is much easier to find out how much copper dissolves than how much is deposited. By measuring the loss in mass of copper at the anode, you can effectively work out how much copper is deposited at the cathode. You will need to measure the quantity of electricity you use:

quantity of electricity in coulombs (C) = Current (A) × time (s)

(96 500 C = 1 faraday (F))

Apparatus

two copper electrodes

power supply

rheostat

ammeter

stopwatch

electrical leads

crocodile clips

beaker

copper(II) sulfate solution

paper towel

access to a balance measuring to 2 decimal places

eye protection

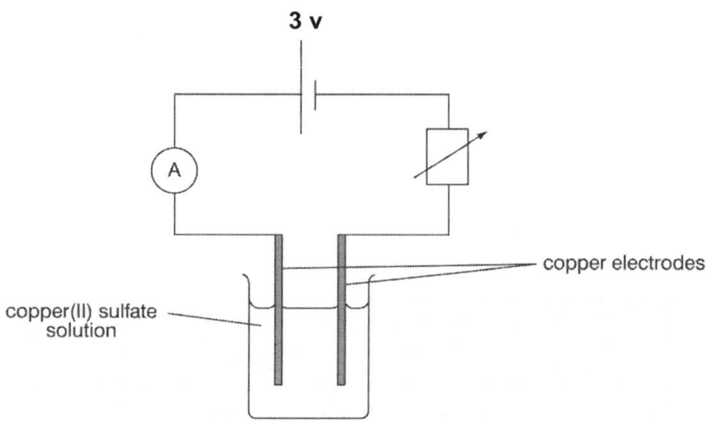

SAFETY INFORMATION

Wear eye protection.

Method

1. Put on your eye protection.
2. Set up the circuit as shown in the diagram above.
3. Clean and dry the copper electrodes, using sand paper if necessary. Weigh one of the electrodes, record the result and connect it in the circuit as the anode.
4. Switch on the circuit and start the stopwatch. Use the rheostat to maintain the current at a constant value. Record the value of the current. Leave the current flowing for about 20 minutes.
5. Stop the stopwatch and record the time. Carefully dry the anode by dabbing it onto a paper towel. Reweigh the anode and record the result.

Make observations and measurements

1. Record your results in the table.

Original mass of copper anode	g
Final mass of copper anode	g
Loss in mass of copper anode	g
Current	A
Time	s

Analyse and interpret data

(A_r Cu = 63.5)

1. Calculate the quantity of electricity you used:

Quantity of electricity = _______ A × _______ s = ____________ coulombs

2. Complete the calculation below:

______________ C produced a loss in mass of __________________ g of copper

______________ C would therefore produce a loss in mass of 1 g of copper

______________ C would therefore produce a loss in mass of 63.5 g of copper

______________ F would therefore produce a loss in mass of 63.5 g of copper

So, _______ F of electricity would be needed to **deposit** 63.5 g of copper at the cathode

3. Copper forms Cu^{2+} ions.

a) Write a half equation showing how copper ions form copper atoms at the cathode.

b) How many faradays of electricity should be needed to produce 1 mole of copper (that is, the theoretical result)?

Evaluate data and methods

1. Your experimental result is likely to be different from the theoretical result. Make a list of possible sources of error in the experiment.

Page 2 of 2

Activity C1i.5 Consolidation and summary

Learning objectives

- To review the learning points of the topic
- To test understanding through answering questions

Learning outcomes

- Be familiar with the knowledge and understanding summarised in the End of Topic Checklist
- Be able to apply this knowledge and understanding by answering the End of Topic Questions

Resources

Student Book pages 99–101

Approach

Use the *Science in context* feature on electroplating in the Student Book page 99 to revise the ideas encountered in the topic.

Refer to the End of Topic Checklist and then ask students to tackle the End of Topic Questions. Some of the concepts encountered here, for example the introduction of half-equations, are quite difficult.

Reassure students that they will have opportunities to revisit these ideas when studying the manufacture of some important chemical – aluminium, sodium hydroxide and chlorine.

Answers

Page 99

1. a) Diagram is the same as Fig. 1.87 but with the teaspoon replaced by a ring.

   b) Anode $Cu(s) \rightarrow Cu^{2+}(aq) + 2e^-$; cathode $Cu^{2+}(aq) + 2e^- \rightarrow Cu(s)$

2. a) 1.6 A × (10 × 60 s) = 960 coulombs

   Number of faradays = 960/96 000 = 0.01

   Ag^+ ion has a charge of +1 (1 faraday will give 1 mole of Ag) so 0.01 faraday gives 1.08 g Ag

   so 1 faraday → 1.08/0.01 = 108 (relative atomic mass Ag)

   b) $Ag^+(aq) + e^- \rightarrow Ag(s)$

   Reduction is gain of electrons. Silver ions have gained electrons to become silver metal.

End of Topic Questions mark scheme

Question	Correct answer	Marks
1 a)	Electrolysis is the breaking down of a (ionic) compound by a flow of electricity.	1 mark
1 b)	An electrolyte is a substance which when molten or dissolved in water allows an electric current to pass through it.	1 mark
1 c)	An electrode is the substance that is used to pass the electric current into and out of the electrolyte during electrolysis.	1 mark
1 d)	The anode is the positively charged electrode.	1 mark
1 e)	The cathode is the negatively charged electrode.	1 mark

Question	Correct answer	Marks
2	The zinc bromide contains ions that are held in a lattice by strong electrostatic attractive forces. In the solid these ions are not free to move and carry the current.	1 mark 1 mark
3	Completed table as shown	4 marks

Electrolyte	Product at the anode	Product at the cathode
Silver bromide	*bromine*	*silver*
Lead(II) chloride	*chlorine*	*lead*
Aluminium oxide	*oxygen*	*aluminium*
Magnesium iodide	iodine	magnesium

4 a)	Anode: chlorine Cathode: sodium	1 mark 1 mark
4 b)	Anode: chlorine and oxygen Cathode: hydrogen	1 mark 1 mark
5 a)	$Al^{3+}(l) + 3e^{-} \rightarrow Al(l)$ (1 mark for correct charges, 1 mark for balanced equation)	2 marks
5 b)	$Na^{+}(l) + e^{-} \rightarrow Na(l)$ (1 mark for correct charges, 1 mark for balanced equation)	2 marks
5 c)	$2O^{2-}(l) \rightarrow O_2(g) + 4e^{-}$ (1 mark for correct charges, 1 mark for balanced equation)	2 marks
5 d)	$2Br^{-}(aq) \rightarrow Br_2(l) + 2e^{-}$ (1 mark for correct charges, 1 mark for balanced equation)	2 marks
5 e)	$4OH^{-}(aq) \rightarrow 2H_2O(l) + O_2(g) + 4e^{-}$ (1 mark for correct charges, 1 mark for balanced equation)	2 marks
6 a)	Cathode	1 mark
6 b)	Silver	1 mark
7	$Al^{3+}(l) + 3e^{-} \rightarrow Al(l)$ 3 F 27 g 1 F 9 g	1 mark 1 mark 1 mark
	Total:	30 marks

Section 1: Exam-style questions mark scheme

Question	Correct answer	Marks
2 a)	i) From top to bottom	
	Proton	1 mark
	Electron	1 mark
	Neutron	1 mark
	ii) 6	1 mark
	iii) 12	1 mark
2 b)	The same: the number of protons and electrons	1 mark
	Different: the numbers of neutrons	1 mark
3 a)	metals and non-metals (or non-metals and metals)	2 marks
	gained and lost (or lost and gained)	2 marks
	high and high	2 marks
3 b)	i) Mg 2,8,2	1 mark
	Cl 2,8,7	1 mark
	ii) Mg 2,8	1 mark
	Cl 2,8,8	1 mark
4	Mass of iron(III) oxide = 4.8 g	
	2 moles of $FeSO_4$ produces 1 mole of Fe_2O_3	1 mark
	304 g $FeSO_4$ produces 160 g Fe_2O_3	1 mark
	9.12 g $FeSO_4$ produces 4.8 g Fe_2O_3	1 mark
	volume of SO_3 = 720 cm^3	
	2 moles of $FeSO_4$ produces 1 mole of SO_3	1 mark
	304 g $FeSO_4$ produces 24 000 cm^3 SO_3	1 mark
	9.12 g $FeSO_4$ produces 720 cm^3 SO_3	1 mark
5 a)	There are mobile electrons between the layers of carbon atoms.	1 mark
5 b)	The lead and bromide ions	1 mark
	will not be able to move when lead(II) bromide is a solid.	1 mark
5 c)	First row: cathode (B)	1 mark
	reduction	1 mark
	Second row: anode (A)	1 mark
	oxidation	1 mark
5 d)	i) 0.05 moles lead	1 mark
	0.05 moles bromine	1 mark
	ii) Formula mass of Br_2 = 160	1 mark
	8 g bromine	1 mark

Section 2. Chemistry of the Elements

Contents

a) The Periodic Table

b) Group 1 elements

c) Group 7 elements

d) Oxygen and oxides

e) Hydrogen and water

f) Reactivity series

g) Tests for ions and gases

h) Exam-style questions

Overview of the section

This section covers much of the inorganic chemistry in the course, although further work appears in Section 5 *Chemistry in society*. The section starts with a study of the Periodic Table as a useful way of ordering elements. Next it focuses on the alkali metals in Group 1 and the halogens in Group 7, contrasting the properties of metals and non-metals, and considering two unusually reactive groups of elements.

The properties of oxygen and hydrogen are discussed next, paying particular attention to the proportion of oxygen in the air, the classification of oxides, the preparation of carbon dioxide and the formation by combustion and the identification water. Analytical techniques used in the identification of gases, anions and cations are then studied. The final topic deals with the reactivity series of metals.

The Periodic Table topic builds on previous work on atoms and atomic structure, setting the scene for a study of Groups 1 and 7 and the reactivity series. Studying the groups and the elements oxygen and hydrogen provides an opportunity to reinforce ideas on chemical formulae and equations. The topic on tests for ions and gases reinforces ideas on ionic bonding and provides a useful recap on the properties of oxygen, carbon dioxide and hydrogen. The reactivity series gives an excellent introduction to the extraction of metals, which is studied as part of Section 5.

In the suggested teaching sequence, all the topics are included in the first year of the course; they provide a basis for some of the later topics (Physical chemistry and Chemistry in society). As with other sections, it is important that the 'orientation' stage gives you an opportunity to establish the extent of your students' prior knowledge.

Starting points

The Student Book provides a 'section opener', a double page spread which sets the scene for the students (see pages 106–107). It is structured in the following way:

- an overview providing details of the areas of study
- six questions, the first three relating to aspects of chemistry that should be reasonably familiar to most students, and the final three relating directly to some of the topics to be studied
- a list of section contents.

The questions are intended to provide a structure for introducing the section.

1. What is an element? How would you define the term?
2. What does the atomic number tell you about its structure?
3. In terms of electronic structures, what is the difference between a metal and a non-metal?

4. You will be learning about the Periodic Table of elements. Look at the Periodic Table and make a list of things that you notice about it.
5. You will be learning about the composition of gases in the air. Do you know, or can you find out, which is the most abundant gas in the air?
6. Make a list of about six to eight metals that you are familiar with. Which metal in your list do you think is the most reactive? Which metal do you think is the least reactive? Explain your choices.

The section opener has two main purposes:

- to acknowledge students' prior learning and to value it
- to provide a benchmark against which future learning can be compared.

The six questions or 'starting points' in the Student Book can be used in a number of different ways to introduce the section:

- You could ask students to consider the questions as an introductory homework task.
- You could put students into groups to share their own ideas and understanding and then report back to the whole class.
- Students could be given access to the Internet, preferably with a tight timescale, to find out the information required.

You could then use a spider chart or other form of wall chart to summarise everybody's ideas.

Recording these initial ideas in this way allows you to retain them for reference as the individual topics are developed. In this way, your students' progress in learning can be readily acknowledged.

C2a The Periodic Table

Introduction

This topic builds on students' understanding of atomic structure and their knowledge of an element as a substance that cannot be broken down by chemical means. The structure of the Periodic Table is introduced as a very useful way to arrange the elements and to separate metals from non-metals.

The properties of metals and non-metals are identified as a result of some straightforward practical activities, which emphasise the investigative skill of *make observations and measurements*. The topic is the basis for more detailed study of Groups 1 and 7 and of metals in the context of the reactivity series.

Links to other topics

Sections	Essential background knowledge	Useful links
1 Principles of chemistry	1b Atoms 1c Atomic structure	1f Ionic compounds 1g Covalent substances 1h Metallic crystals
2 Chemistry of the elements		2b Group 1 elements 2c Group 7 elements 2d Oxygen and oxides 2f Reactivity series

Topic overview

C2a.1	**Orientation** The purpose of this short activity is to recap students' understanding of the nature of elements and atomic structure. You may use demonstrations to set the scene for this topic and some of the subsequent topics on particular groups of elements.
C2a.2	**Arrangement of the elements** This activity concentrates on the structure of the Periodic Table, the use of the atomic number to order the elements and how they are arranged in groups and periods. It also covers the link between the position of an element in the Periodic Table and its likely metal/non-metal classification and electron arrangement. The extent to which you emphasise ideas on ion charges will depend on whether you have already covered Topic 1f *Ionic compounds*.
C2a.3	**Characteristic properties of metals and non-metals** This activity focuses on the characteristic physical properties of metal and non-metals and then the classification of oxides as basic or acidic. Two straightforward practical activities can be used to emphasise these differences between metals and non-metals.
C2a.4	**Consolidation and summary** This activity provides an opportunity for a quick recap on the ideas encountered in the topic and time for the students to answer the End of Topic Questions in the Student Book.

Activity C2a.1 Orientation

Learning objectives

- To revise understanding of what an element is
- To revise knowledge of atomic structure

Learning outcomes

- Know that an element is a substance that cannot be broken down into anything simpler by chemical means
- Known how atomic structure is related to the atomic number
- Know how electrons are arranged in shells around the nucleus

Resources

Student Book page 108

CD–ROM: C2a.1a_tech_notes

Resources for demonstration (see below)

Approach

This introductory activity can be very short, particularly if students show a good knowledge and understanding of previous work on atoms and atomic structure. You could base your approach on a quick quiz of 10 questions to probe their understanding of the nature of an element (as opposed to a compound or a mixture) and the arrangement of protons, neutrons and electrons in an atom.

You can then justify the study of the Periodic Table by a quick demonstration of the nature of three elements: sodium (symbol Na, a shiny, soft element stored under oil); copper (symbol Cu, a hard, shiny metal with high melting point); chlorine (symbol Cl, a pale yellow gas).

These elements are clearly very different from each other. Where are they found in the Periodic Table? Perhaps there is a clear link between the properties of different elements and their position in the Periodic Table? Perhaps the Periodic Table as an arrangement of elements might be a really useful way of simplifying the study of all these elements?

Technician's notes

You will need the following resources for a demonstration.

Be sure to check the latest safety notes on these resources before proceeding.

bottle of sodium
copper foil
gas jar of chlorine (should be kept in a fume cupboard)
sharp knife, tongs and filter paper

Activity C2a.2 The arrangement of the elements

Learning objectives

- To look in more detail at the arrangement of elements in the Periodic Table
- To emphasise the positions of metals/non-metals/metalloids; trends in reactivity; electron arrangements and ion charges

Learning outcomes

- Know that metals are found on the left-hand side and in the middle of the Periodic Table
- Know that non-metals are found on the right-hand side of the Periodic Table
- Understand that the chemical properties of an element are largely governed by the number of electrons in the outermost electron shell
- Know that in groups of metals reactivity increases down the group whereas in groups of non-metals reactivity decreases down the group
- Understand that the noble gases in Group 0 are very unreactive because they have full outer shells or 8 electrons in their outer shells

Common misconceptions

Some students think that ions have full outer electron shells – in fact they have the electron arrangement of the nearest noble gas (or have full electron shells or 8 electrons in the outer shell).

Resources

Student Book pages 108–110

Approach

Refer briefly to the historical development of the Periodic Table and the work of Mendeleev, who is generally credited with being the leading originator of the modern Periodic Table. Emphasise the following aspects of the arrangement of the table:

1. The labelling of groups and periods

2. The relative positions of metals, non-metals and metalloids

3. Trends in reactivity for metals and non-metals. The Student Book explains these trends, based on an understanding of the formation of ions. For metals, the further the outer electrons are from the attractive force of the nucleus the more readily the atom will lose an electron and form a positive ion; for non-metals, the smaller the atom the more an 'incoming' electron will experience the attractive force of the nucleus and so the more readily the atom will form a negative ion

4. The relationship between group number and ion charge.

Answers

Page 110

1. a) 20

   b) The atomic number is the number of protons (which equals the number of electrons) in an atom of the element. Calcium has 20 protons and 20 electrons.

   c) Group 2 d) Period 4 e) Calcium is a metal

2. Halogens

3. Halogens are non-metals.

Activity C2a.3 Characteristic properties of metals and non-metals

Learning objectives

- To explore the physical and chemical properties of a range of metals and non-metals
- To identify the characteristic properties of metals and non-metals

Learning outcomes

- Be able to explain that metals conduct electricity and form basic oxides
- Be able to explain that non-metals are generally non-conductors of electricity and usually form acidic oxides

Common misconceptions

All metals are magnetic. There are only three – iron, nickel and cobalt.

Resources

Student Book pages 110–113

Worksheet C2a.3a Physical properties of metals and non-metals

Worksheet C2a.3b Investigating oxides

Files on CD–ROM: C2a.3a_worksheet, C2a.3b_worksheet, C2a.3a_tech_notes, C2a.3b_tech_notes

Resources for a class practical (see below)

Approach

The two worksheets give students an opportunity to explore the physical properties of some elements as well as the chemical nature of the oxides of metals and non-metals.

Worksheet C2.3a will require a teacher demonstration to compare the conduction of heat by the different elements.

The characteristic physical properties of metals and non-metals are summarised in the Student Book on pages 110 and 111.

Worksheet C2a.3b is a short practical activity. Students are likely to classify aluminium oxide and silicon dioxide as neutral oxides. It would be helpful if you extend the definition of a basic oxide as one that also 'reacts with acids' and for an acidic oxide 'reacts with alkalis'. With this expanded definition, aluminium oxide can be classified as a basic oxide and silicon oxide as an acidic oxide.

Technician's notes

You will need the following resources for a demonstration and class practical.

Be sure to check the latest safety notes on these resources before proceeding.

Physical properties

samples (strips – 3 cm × 1cm – where possible) of aluminium, carbon, copper, iron, lead, sulfur, zinc
electrical circuit – power supply (6 V), lamp, leads, crocodile clips
Bunsen burner, tripod, gauze (or access to hotplate), longer strips of aluminium, carbon, copper, iron, zinc (all the same length), candle wax or vaseline
eye protection

Chemical properties

sodium hydroxide 0.1M (labelled ‘sodium oxide (aq)’)
phosphoric acid 0.1M (labelled ‘phosphorus oxide (aq)’)
sulfuric acid 0.1M (labelled ‘sulfur oxide (aq)’)
samples of magnesium oxide, aluminium oxide and silicon oxide (sand)
test tubes, universal indicator and pH chart
eye protection

Answers

Page 112

1. A metalloid has some properties typical of a metal and some of a non-metal.
2. The metal can be drawn into a wire.
3. The metal can be beaten into a sheet.
4. Carbon in the form of graphite will conduct electricity.
5. Non-metals usually form acidic oxides.

Page 113

1. Aluminium has 3 electrons in the outer shell.
2. Oxygen will form an ion (the oxide ion) with a 2- charge.
3. Fluorine (F)
4. Barium (Ba)

C2a.3a Physical properties of metals and non-metals

Your task is to identify the characteristic physical properties of metals and non-metals.

Apparatus

Samples of the following elements:

aluminium, carbon, copper, iron, lead, sulfur and zinc

electrical circuit (as shown in the diagram)

magnet

Bunsen burner and heat-proof mat

eye protection

SAFETY INFORMATION

Wash hands after handling lead.

Wear eye protection.

Method

1. You will be provided with a small sample of the following elements: aluminium, carbon, copper, iron, lead, sulfur, zinc. By observation and testing record your results in the table below.

To test for conduction of heat

2. Your teacher will demonstrate a simple way of comparing the rate of conduction of heat. Equal lengths of the elements will have a drawing pin attached to one end by some candle wax or Vaseline. The other ends of the elements will be heated on a hotplate or in a Bunsen burner flame. As heat is conducted down the element the wax/ Vaseline will soften and the drawing pin will fall off. Lead and sulfur cannot be tested in this way (the lead will melt and the sulfur will catch fire).

To test for conduction of electricity

3. Use the circuit shown in the diagram below and test each element in turn.

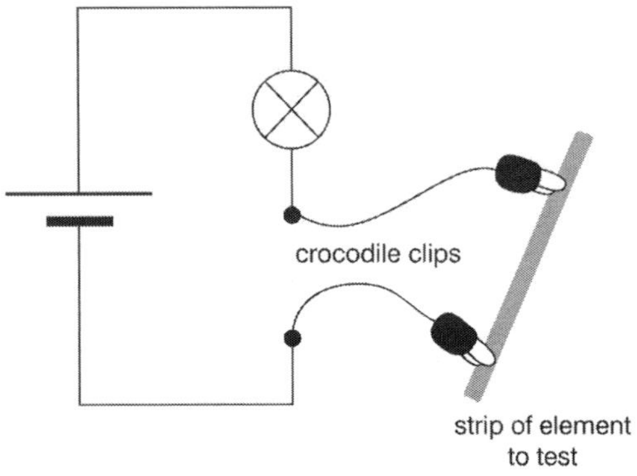

Page 1 of 2

Make observations and measurements

Element	Metal or non-metal	Appearance	Dull or shiny	Flexible or brittle	Good/ poor conductor of heat	Good/ poor conductor of electricity	Magnetic/ non-magnetic
aluminium							
carbon							
copper							
iron							
lead							
sulfur							
zinc							

Analyse and interpret data

4. Use your results to complete this summary chart.

Physical property	Metal	Non-metal
Appearance		
Conduction of heat		
Conduction of electricity		
Magnetism		

5. List any exceptions to the conclusions you have drawn.

Page 2 of 2

C2a.3b Investigating oxides

Your task is to test the pH of a range of metal and non-metallic oxides and to identify any patterns in your results. Oxides which form alkaline solutions are known as **basic oxides**; oxides which form acidic solutions are known as **acidic oxides**. Some oxides when tested appear to be **neutral oxides** but this may be because they do not dissolve in water.

Apparatus

samples of magnesium oxide, aluminium oxide and silicon oxide

samples of sodium oxide/hydroxide (aq), phosphorus oxide (aq), sulfur oxide (aq)

test tubes, universal indicator and pH chart

eye goggles

SAFETY INFORMATION

Sodium oxide (aq) – sodium hydroxide is an irritant.

Wear eye goggles.

Carry out in a fume cupboard.

Method

1. Put on your eye protection.
2. You will have access to samples of the oxides given in the table. Some will already be in solution, others will be solids.
3. To test the solutions use universal indicator paper and the colour chart to determine the pH of the solution.
4. To test the solids, add a small amount of each to some distilled water in a test tube and stir to try to dissolve the solid before testing with universal indicator paper.

Page 1 of 2

Make observations and measurements

1. Record your results in the table.

	Sodium oxide (in solution exists as sodium hydroxide)	**Magnesium oxide**	**Aluminium oxide**	**Silicon oxide**	**Phosphorus oxide**	**Sulfur oxide**
Metal oxide or non-metal oxide						
State of matter	aq	s	s	s	aq	aq
pH of solution						
Acidic or neutral or basic (alkaline) oxide						

Analyse and interpret data

1. What pattern can you identify in your results?
2. Are there any exceptions to this pattern?

Activity C2a.4 Consolidation and summary

Learning objectives

- To review the learning points of the topic
- To test understanding through answering questions

Learning outcomes

- Be familiar with the knowledge and understanding summarised in the End of Topic Checklist
- Be able to apply this knowledge and understanding by answering End of Topic Questions in the Student Book

Resources

Student Book pages 114–115

Approach

This activity can be very short with the main focus being on the End of Topic Questions. Ideas on trends in reactivity will feature strongly in topics on Group 1 and Group 7 and in the topic on the *Reactivity series*.

Answers

End of Topic Questions mark scheme

Question	Correct answer	Marks
1 a)	b	1 mark
1 b)	a	1 mark
1 c)	d	1 mark
1 d)	c	1 mark
1 e)	b, d and f	1 mark
1 f)	d	1 mark
2 a)	Metal	1 mark
2 b)	Basic oxide – metals form basic oxides.	1 mark
3 a)	i) Can be beaten into sheets/shape ii) Can be drawn into wires	1 mark 1 mark
3 b)	Any metal (ideally not Group 1 but allow)	1 mark
4	Elements in the same group have the same number of electrons in their outer electron shell.	1 mark
5 a)	Na^+	1 mark
5 b)	Cl^-	1 mark
6 a)	Reactivity increases as you go down a group of metals.	1 mark
6 b)	Reactivity decreases down a group of non-metals.	1 mark
7	The noble gases have a full outer shell of electrons or 8 electrons in their outer electron shells. They do not easily gain or lose electrons.	1 mark 1 mark
	Total:	18 marks

C2b Group 1 elements

Introduction

This topic follows on naturally from Topic 2a *The Periodic Table*. Students will know that Group 1 is a group of metals and will probably know that sodium is an unusual metal, as it is stored under oil in the laboratory. The topic is very short and the first two activities could easily be covered in the same lesson.

The topic centres on the reactions of the Group 1 elements with water. The Student Book also covers reactions of Group 1 elements with chlorine to illustrate their high reactivity, although this is not required by the course specification. Alkali metals have obvious links with Topic 4a *Acids, alkalis and salts* and the manufacture of sodium hydroxide in Topic 5d *The industrial manufacture of chemicals*.

Links to other topics

Sections	Essential background knowledge	Useful links
1 Principles of chemistry	1e Chemical formulae and chemical equations	1f Ionic compounds
2 Chemistry of the elements	2a The Periodic Table	2f Reactivity series
4 Physical chemistry		4a Acids, alkalis and salts
5 Chemistry in society		5d The industrial manufacture of chemicals

Topic overview

C2b.1	**Orientation** This is a very short activity re-emphasising the position of the Group 1 elements in the Periodic Table and introducing the elements lithium, sodium and potassium. These are very unusual metals, as they are soft and stored under oil in the laboratory. Equations for the reaction of the elements with oxygen are introduced.
C2b.2	**Reactions of the group 1 elements** This activity involves demonstrating the reactions of lithium, sodium and potassium with water. It emphasises the formation of hydrogen gas and an alkaline solution. Equations for the reactions are developed.
C2b.3	**Consolidation and summary** This is another short activity which explains the trend in reactivity in the Group 1 elements. It also provides an opportunity for a quick recap on learning points, as well as gives students time to answer the End of Topic Questions in the Student Book.

Activity C2b.1 Orientation

Learning objectives

- To revise the nature of the grouping of elements in the Periodic Table
- To establish Group 1 as a 'family' of very reactive metals

Learning outcomes

- Know that lithium, sodium and potassium are soft, reactive metals and stored in oil

Resources

Student Book pages 116–117

Files on CD–ROM: C2b.2a_tech_notes

Resources for a demonstration (see below)

Approach

Quickly recap on previous work on the Periodic Table. Ask students what they already know about the Group 1 elements. Show samples of the elements in bottles. Ask students why the elements are stored in oil. Construct equations for the reactions of the metals with oxygen (from the air).

Technician's notes

You will need the following resources for a demonstration.

Be sure to check the latest safety notes on these resources before proceeding.

lithium
sodium
potassium
filter paper and knife

Answers

Page 117

1. They react with water to form alkaline solutions.

2. 1 electron in the outer shell.

3. The potassium atom is larger than the lithium atom so the outer electron is further from the attraction of the nucleus and can be more easily removed.

4. They are soft to cut (also have very low melting points).

Activity C2b.2 Reactions of the Group 1 elements

Learning objectives

- To demonstrate the reactions of lithium, sodium and potassium with water
- To write balanced chemical equations to represent these reactions

Learning outcomes

- Be able to describe the reactions of lithium, sodium and potassium with water
- Know that in their reactions with water an alkali and hydrogen are produced
- Know that the elements become increasingly reactive as you go down the group, from lithium to francium

Resources

Student Book page 118–121

Files CD–ROM: C2b.2a_tech_notes

Resources for a demonstration (see below)

Approach

Demonstrate the reaction of lithium, sodium and potassium with water. The oil must be wiped off the metal once it has been cut. A small trough of water on an overhead projector provides good visibility. Alternatively, you may use a large trough of water.

Use a safety screen and carefully control the size of the metal samples. In a small trough of water, use samples no bigger than 1 mm × 1 mm; in a large trough of water the metal samples must be no bigger than 2 mm × 2 mm. Demonstrate the reactions in turn, using fresh water each time. Add universal indicator to the solution to illustrate how an alkaline solution forms – hence the family name 'alkali metals'.

The increased reactivity down the group should be easy to observe from lithium (which doesn't usually melt) to sodium (which melts into a sphere instantaneously) to potassium (which melts and catches fire on contact with the water).

There are many videos clips available on the Internet of rubidium reacting with water, which can form a stimulating conclusion to the demonstrations. Ask students to predict what will happen before they watch the clip.

More detailed observations are provided in the Student Book. Introduce equations for the reactions.

Technician's notes

You will need the following resources for a demonstration.

Be sure to check the latest safety notes on these resources before proceeding.

lithium
sodium
potassium
filter paper and knife
glass bowl
overhead projector and safety screen
eye protection

Answers

Page 119

1. Sodium oxide is white.
2. Hydrogen. The solution formed is potassium hydroxide.
3. The compounds are soluble.

Page 121

1. a) A period is a horizontal row of elements in the Periodic Table. All the elements in the same period have the same number of electron shells.

For example: Period 1 elements: hydrogen (1) to helium (2) – 1 electron shell

Period 2 elements: lithium (2,1) to neon (2,8) – 2 electron shells

Period 3 elements sodium (2,8,1) to argon (2,8,8) – 3 electron shells

b) sodium, magnesium, aluminium, silicon, phosphorus, sulfur, chlorine, argon

c) 2,8,1; 2,8,2; 2,8,3; 2,8,4; 2,8,5; 2,8,6; 2,8,7; 2,8,8

d) First element 2,8,1 (1 electron in the outer shell, easily lost, will form positive ions, reactive metal)

Last element 2,8,8. (8 electrons in outer shell, full outer shell, noble gas, unreactive)

2. a) A group is a vertical column of elements having similar chemical properties because of their outer shell electronic structure.

b) lithium, sodium, potassium

c) All the elements in this group 1 have one electron in the outer shell.

d) The reactivity of these elements depends upon the ease at which the outer electrons are lost. One electron can easily be lost to form positive ions. The ease at which it can be lost increases down the group as the electron is less tightly held in the atom and therefore reactivity increases down the group.

Activity C2b.3 Consolidation and summary

Learning objectives

- To provide an explanation for the trend in reactivity of the Group 1 elements
- To review the learning points of the topic
- To test understanding through answering questions

Learning outcomes

- Be able to explain that the reactivity of the elements depends on the distance between the outer electron and the nucleus; the further the distance of the outer electron from the nucleus, the more reactive the metal is
- Be familiar with the knowledge and understanding summarised in the End of Topic Checklist
- Be able to apply this knowledge and understanding by answering End of Topic Questions in the Student Book

Resources

Student Book pages 122–123

Approach

Provide an explanation for the trend in reactivity of lithium, sodium and potassium – students should have seen this explanation in Topic 2a *The Periodic Table* topic. Focus on the Checklist and End of Topic Questions in the Student Book to recap on the learning points of the topic. The work should be very fresh in a short topic like this, so this activity should be completed quickly.

Answers

End of Topic Questions mark scheme

Question	Correct answer	Marks
1 a)	Potassium	1 mark
1 b)	To prevent reaction with air (oxygen) or water	1 mark
1 c)	Potassium	1 mark
1 d)	The elements react with the oxygen in the air.	1 mark
1 e)	Sodium is less dense than water.	1 mark
2	The group 1 elements react with water to form alkalis. The alkalis are hydroxides, such as sodium hydroxide.	1 mark 1 mark
3 a)	lithium + oxygen → lithium oxide $4Li(s) + O_2(g) \rightarrow 2Li_2O(s)$ (1 mark for correct formulae, 1 mark for balancing)	1 mark 2 marks
3 b)	potassium + water → potassium hydroxide + hydrogen $2K(s) + 2H_2O(l) \rightarrow 2KOH(aq) + H_2(g)$ (1 mark for correct formulae, 1 mark for balancing)	1 mark 2 marks
3 c)	potassium + chlorine → potassium chloride $2K(s) + Cl_2(g) \rightarrow 2KCl(s)$ (1 mark for correct formulae, 1 mark for balancing)	1 mark 2 marks
4	Potassium is a larger atom than sodium. The outer electron in potassium is further away from the nucleus than the outer electron in sodium. The outer electron in potassium is easier to remove (as in a reaction) than the outer electron in sodium.	1 mark 1 mark 1 mark
5 a)	Rubidium is a solid.	1 mark
5 b)	i) Hydrogen. ii) Rubidium hydroxide would be formed. Universal indicator would show the presence of an alkali.	1 mark 1 mark 1 mark
5 c)	Rubidium would be more reactive than potassium. The outer electron in rubidium is further away from the nucleus than that in potassium and so it will be more easily transferred in a reaction.	1 mark 1 mark
	Total:	25 marks

C2c Group 7 elements

Introduction

This is a short topic that is likely to follow on from Topic 2a *The Periodic Table* and Topic 2b *Group 1 elements*. Students should be familiar with the position of non-metals in the Periodic Table, so they should recognise Group 7 as non-metals.

The topic concentrates on the appearance of the three common halogens (chlorine, bromine and iodine), their relative reactivities and how these are confirmed by displacement reactions involving the halogen and solution of their salts. One activity looks at hydrogen chloride and the effect of water as a solvent when the gas is dissolved. This links to Topic 4a *Acids, alkalis and salts*. The industrial manufacture of chlorine is covered in Topic 5d *The industrial manufacture of chemicals*. There is an opportunity for some student practical work on displacement reactions.

Links to other topics

Sections	Essential background knowledge	Useful links
1 Principles of chemistry	1e Chemical formulae and equations	1f Ionic compounds 1g Covalent substances
2 Chemistry of the elements	2a The Periodic Table	2g Tests for ions and gases
4 Physical chemistry		4a Acids, alkalis and salts
5 Chemistry in society		5d The industrial manufacture of chemicals

Topic overview

C2c.1	**Orientation** This activity very briefly recaps on the structure of the Periodic Table, followed by an introduction to the elements of chlorine, bromine and iodine. Fluorine and astatine are briefly mentioned in the context of predicting their physical states and relative reactivities.
C2c.2	**Displacement reactions of the halogens** This activity focuses on the potential reactions of chlorine, bromine and iodine with solutions of their salts. You will investigate the redox nature of these reactions, using the definitions of oxidation and reduction in terms of electron loss/gain.
C2c.3	**Hydrogen chloride** Hydrogen chloride is introduced as an important compound made from chlorine. The importance of water as the solvent in hydrochloric acid in producing $H^+(aq)$ ions is emphasised and compared with solutions of hydrogen chloride with some organic solvents, such as methylbenzene.
C2c.4	**Consolidation and summary** This activity allows a quick recap on the ideas encountered in the topic, particularly displacement reactions, and an opportunity for students to answer the End of Topic Questions in the Student Book.

Activity C2c.1 Orientation

Learning objectives

- To review the significance of the positioning of the Group 7 elements in the Periodic Table and the trend in reactivity for a group of non-metals
- To study the appearances of the Group 7 elements

Learning outcomes

- Know the colours and physical states of chlorine, bromine and iodine at room temperature
- Know the relative reactivity of the Group 7 elements
- Be able to predict the likely properties of fluorine and astatine

Resources

Student Book pages 124–125

Resources for a demonstration (see below)

Files on CD-ROM: C2c.1a_tech_notes

Approach

Ask students to tell you what they already know about the Group 7 elements – they should know that they are non-metals, that they all have 7 electrons in their outer electron shells and that reactivity (for non-metals) decreases as you descend the group. Explain that the elements have the family name 'halogens' which means 'salt maker', in other words, these elements react with metals to form salts (such as chlorides, bromides and iodides).

Show students samples of the three halogens: chlorine is a pale green gas; bromine is a red brown liquid and iodine is a black solid. Explain that because chlorine and bromine are toxic in the laboratory they are often used in solution – chlorine solution is colourless; bromine solution is orange. Put the three elements in order of reactivity and then put cards for fluorine and astatine in their respective positions.

Ask students to predict the appearance and reactivity of these other two elements (fluorine – a pale yellow gas; astatine a black solid). Refer the students to the feature on fluorine on page 130 of the Student Book.

Finally, explain the trend in reactivity: the smallest atom most easily gains an electron to achieve the electron configuration of the nearest noble gas (the 'incoming' electron would feel a greater attraction from the nucleus).

Technician's notes

You will need the following resources for a demonstration.

Be sure to check the latest safety notes on these resources before proceeding.

a gas jar of chlorine (in a fume cupboard)
bromine liquid in a bottle/gas jar
iodine solid (0.25g)

These should all be stored in a fume cupboard.

Activity C2c.2 Displacement reactions of the halogens

Learning objectives

- To consider the nature of the displacement reactions involving the halogen elements
- To introduce or revise the concept of redox reactions

Learning outcomes

- Be able to describe and explain displacement reactions involving the halogens and solutions of their salts
- Understand that displacement reactions are redox reactions and involve oxidation (loss of electrons) and reduction (gain of electrons)

Resources

Student Book page 126–129

Worksheet C2c.2a Displacement reactions of halogens

Files on CD-ROM: C2c.2a_worksheet, C2c.2a_tech_notes

Resources for a class practical (see below)

Approach

Explain displacement reactions in terms of the halogen atoms competing for a metal. Students can then undertake practical work using Worksheet C2c.2a. Explain that halogens are used in solution to reduce the risks posed by these chemicals. The cyclohexane is used to distinguish more precisely between solutions containing bromine and those containing iodine. You will need to spend time on developing the equations for the reactions.

The Student Book explains the redox nature of these reactions with an ionic equation and then with half-equations. Ionic equations are not a specific requirement of this course, but half-equations are, so you will need to decide the best way to approach this. Students are expected to understand that reduction is the gain of electrons and oxidation is the loss of electrons.

Developing Investigative Skills on page 128 of the Student Book deals with displacement reactions. You could use this as a consolidation after the practical activity or leave it for the consolidation and summary activity at the end of the topic.

Technician's notes

You will need the following resources for a class practical.

Be sure to check the latest safety notes on these resources before proceeding.

chlorine solution (2% maximum)
bromine solution (1% or 0.06M)
iodine solution (0.1M approx.)
0.1M solution sodium chloride
0.1M solution sodium bromide
0.1M solution sodium iodide
cyclohexane
test tubes, droppers, glass rods
eye protection

Answers

Page 127

1. 7 electrons in the outer shell.

2. The atoms only need to gain one electron to achieve 8 in the outer shell.

3. The chlorine molecule is made up of two atoms combined/bonded together, Cl_2.

4. A displacement reaction is where one Group 7 element takes the place of another in a metallic compound.

5. The displacement reaction involves one Group 7 element being reduced (gaining electrons) and one being oxidised (losing electrons).

Developing investigative skills, page 128

1. a) Make sure there are no naked flames in the room.

   b) Work in a fume cupboard or make sure the room is well ventilated.

   c) Don't dispose of the liquid mixture containing the cyclohexane down the sink.

2. Chlorine is more reactive than bromine (it displaced the bromine in sodium bromide).

   Chlorine is more reactive than iodine (it displaced the iodine in sodium iodide).

3. $Br_2(aq) + 2NaI(aq) \rightarrow 2NaBr(aq) + I_2(aq)$

4.Tube 6. Iodine is less reactive than bromine and so cannot displace bromine from sodium bromide. (The cyclohexane layer should have been violet.)

C2c.2a Displacement reactions of the halogens

Under the right conditions, the halogen elements will compete with each other to form compounds with metals. For example, if bromine is added to sodium iodide solution there are two possibilities:

1. Bromine solution + sodium iodide solution → no change
2. Bromine solution + sodium iodide solution → sodium bromide solution + iodine solution

If as in 1 there is no change, then iodine is more reactive than bromine.

If the reaction in 2 takes place then bromine is more reactive than iodine.

From your knowledge of the trend in reactivity of the halogen elements, you should be able to predict which more likely, 1 or 2.

Apparatus

solutions of chlorine, bromine and iodine

solutions of sodium chloride, sodium bromide and sodium iodide

cyclohexane

test tubes, glass rod

eye protection

SAFETY INFORMATION

Cyclohexane is highly flammable.

Chlorine solution, bromine solution, iodine solution are all irritants.

Wear eye protection.

Method

Part 1

1. Put on your eye protection.
2. Put samples of chlorine solution, bromine solution and iodine solution in separate test tubes.
3. Add 1 cm depth of cyclohexane to each tube and carefully stir with a glass rod.
4. Record the colour of the cyclohexane layer in the table.

CARE: Do not pour cyclohexane mixtures into the sink.
Use the beaker provided by your teacher.

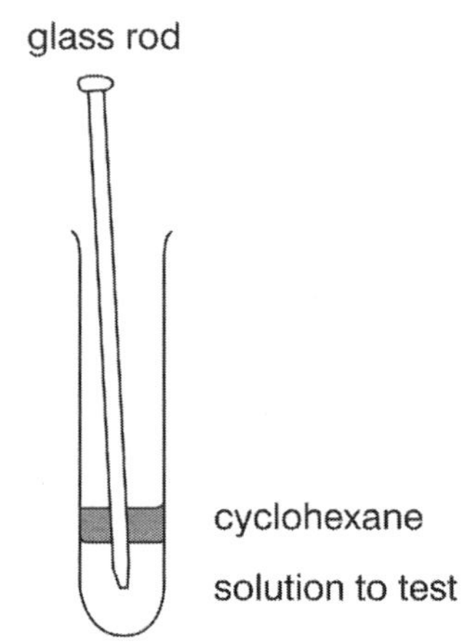

Page 1 of 2

Part 2

4. Now carry out the tests shown in Table 2.

5. Put about 0.5 cm depth of solution A in a test tube. Add about 1 cm depth of solution B. Finally add about 1 cm depth of cyclohexane and stir the mixture with a glass rod.

6. Record the colour of the cyclohexane layer in the table.

Make observations and measurements

Table 1

Solution	Colour of cyclohexane layer
chlorine	
bromine	
iodine	

Table 2

Solution A	Solution B	Colour of cyclohexane layer	Which halogen remains in the cyclohexane layer?
Chlorine	sodium bromide		
Chlorine	sodium iodide		
Bromine	sodium chloride		
Bromine	sodium iodide		
Iodine	sodium chloride		
Iodine	sodium bromide		

Analyse and interpret data

7. Write word equations for any reactions that occurred.

8. What is the order of reactivity of the halogen elements as indicated by these displacement reactions?

9. Write fully balanced chemical equations for any reactions that occurred. In each case identify the reactant that has been oxidised and the reactant that has been reduced.

10. Would you expect a reaction between astatine and sodium bromide solution? Explain your answer.

Activity C2c.3 Hydrogen chloride

Learning objectives

- To contrast hydrochloric acid (a solution of hydrogen chloride in water) with a solution of hydrogen chloride in methylbenzene

Learning outcomes

- Understand that hydrogen chloride dissolves in water to form hydrochloric acid which dissociates into hydrogen ($H^+(aq)$) ions and chloride ($Cl^-(aq)$) ions
- Understand that hydrogen chloride in other solvents, such as methylbenzene, does not dissociate into ions and so shows no acidic properties

Resources

Student Book page 129–131

Files on CD–ROM: C2c.3a_tech_notes

Resources for a demonstration (see below)

Approach

This is a short activity. Explain that hydrogen chloride is an important chemical that is made from chlorine and that large quantities of it are used to make hydrochloric acid. If students have studied Topic 4a *Acids, alkalis and salts* they will already know the significance of acidic properties being related to the existence of $H^+(aq)$ ions. If not, introduce this idea now and consolidate it when you are teaching acids in more detail later.

Demonstrate the different conductivities of a solution of hydrogen chloride dissolved in water with a solution of hydrogen chloride dissolved in methylbenzene. Explain that the hydrogen chloride molecule in water dissociates into ions, but this does not happen in a solvent such as methylbenzene.

Technician's notes

You will need the following resources for a demonstration.

Remember to wear eye protection and carry out in a fume cupboard.

Be sure to check the latest safety notes on these resources before proceeding.

hydrochloric acid (0.1M)
hydrogen chloride dissolved in methylbenzene (0.1M approx.)
power supply, lamp, leads, crocodile clips, carbon electrodes

Answers

Page 130 (top)

1. Chlorine kills any bacteria that might be present in the water.
2. H^+ and Cl^- ions are the main ions present in aqueous hydrochloric acid.
3. In methylbenzene no H^+ ions are formed.

Page 130 (bottom)

1. a)

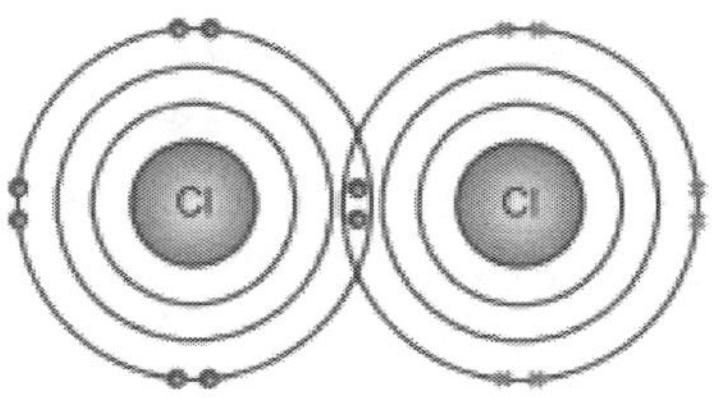

Cl – Cl

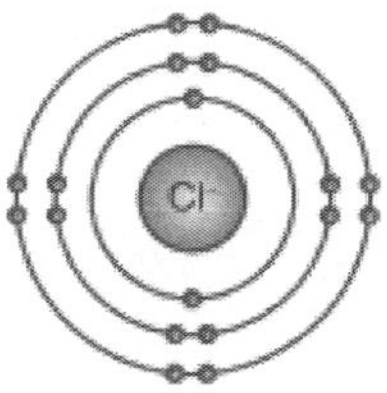

Cl^-

b) $2Cl^-(aq) \rightarrow Cl_2(g) + 2e^-$. The chloride ions lose electrons.

c) $2NaOH(aq) + Cl_2(g) \rightarrow NaCl(aq) + NaClO(aq) + H_2O(l)$

d) Chlorine is a more reactive halogen than bromine. Chlorine will displace bromine from a solution of bromide ions. (Chlorine will oxidise bromide ions to bromine.)

Observations: chlorine water is pale green. When this is added to a colourless solution of potassium bromide the resulting solution will turn orange due to the presence of bromine.)

2. a) The reactivity of fluorine is due to its electronic structure 2,7. Fluorine only needs to gain one electron to form a fluoride ion. This is very easy because of its small size and high attractive force of the nucleus.

b) $F_2(aq) + 2e^- \rightarrow 2F^-(aq)$

c) Both chlorine and iodine are less reactive than fluorine. Fluorine could only be displaced from fluoride ions by a more reactive halogen. As there are no halogens that are more reactive than fluorine, fluorine will not be displaced from fluoride ions.

Activity C2c.4 Consolidation and summary

Learning objectives

- To review the learning points of the topic
- To test understanding through answering questions

Learning outcomes

- Be familiar with the knowledge and understanding summarised in the End of Topic Checklist
- Be able to apply this knowledge and understanding by answering End of Topic Questions

Resources

Student Book pages 132–133

Approach

Revisit displacement reactions and, if not already covered, ask the students to answer the questions related to the *Developing Investigative Skills* feature in the Student Book. Spend some time revising the equations for displacement reactions and the redox nature of these reactions. Focus on the Checklist and End of Topic Questions in the Student Book to recap on the main learning points in the topic.

Answers

End of Topic Questions mark scheme

Question	Correct answer	Marks
1 a)	Chlorine	1 mark
1 b)	Bromine	1 mark
1 c)	Iodine	1 mark
1 d)	Bromine is a red brown liquid.	1 mark
2 a)	The Group 7 elements have 7 electrons in the outer electron shell. The Group 7 elements only need to gain 1 electron in a reaction to achieve a full outer electron shell or 8 electrons in their outer electron shell.	1 mark 1 mark
2 b)	Fluorine is the smallest of the Group 7 atoms and so its outer electrons feel the strongest attraction to the nucleus. Fluorine therefore can gain an electron (in a reaction) more easily than the other elements can.	1 mark 1 mark
3 a)	sodium + chlorine → sodium chloride $2Na(s) + Cl_2(g) \rightarrow 2NaCl(s)$ (1 mark for formulae, 1 mark for balancing)	1 mark 2 marks
3 b)	magnesium + bromine → magnesium bromide $Mg(s) + Br_2(l) \rightarrow MgBr_2(s)$ (1 mark for formulae, 1 mark for balancing)	1 mark 2 marks
3 c)	hydrogen + fluorine → hydrogen fluoride $H_2(g) + F_2(g) \rightarrow 2HF(g)$ (1 mark for formulae, 1 mark for balancing)	1 mark 2 marks
4 a)	This is a displacement reaction.	1 mark
4 b)	$Br_2(l) + 2NaI(aq) \rightarrow 2NaBr(aq) + I_2(aq)$ (1 mark for formulae, 1 mark for balancing)	2 marks
4 c)	I) Oxidation is the loss of electrons. The iodide ions have lost electrons and been oxidised. ii) Reduction is the gain of electrons. The bromine has gained electrons and been reduced.	1 mark 1 mark 1 mark 1 mark
5 a)	The hydrogen ion, H^+.	1 mark
5 b)	In solvents such as methylbenzene the hydrogen chloride does not form H^+ ions.	1 mark
	Total:	26 marks

C2d Oxygen and oxides

Introduction

The concept of basic and acidic oxides was introduced in Topic 2a *The Periodic Table*. This topic starts by considering oxygen as an important component of the air, how it can be prepared in the laboratory and its reactions with some common elements to form oxides. Carbon dioxide as a greenhouse gas and its potential contribution to climate change are particularly important.

The topic gives students opportunities to prepare and test oxygen and carbon dioxide, linking to Topic 2g *Tests for ions and gases*. The formation of carbon dioxide by the oxidation of carbon provides a link to Topics 3b *Alkanes* and 5b *Crude oil*. Students will be able to practise the skills of *demonstrate and describe techniques* and *make observations and measurements*.

Links to other topics

Sections	**Essential background knowledge**	**Useful links**
1 Principles of chemistry	1e Chemical formulae and chemical equations	
2 Chemistry of the elements	2a The Periodic Table	2f Reactivity series 2g Tests for ions and gases
3 Organic chemistry		3b Alkanes
4 Chemistry in society		5b Crude oil

Topic overview

C2d.1	**Orientation** This brief introductory activity sets the context in terms of the position of oxygen in the Periodic Table: a non-metal at the top of its group (hence reactive). You will then introduce the percentage composition of the air, with oxygen accounting for about one fifth of the gases in the air. Reactions of oxygen with magnesium, sulfur and carbon can be demonstrated.
C2d.2	**Determining the percentage composition of oxygen in the air** This activity looks at experimental procedures for confirming the proportion of oxygen in the air at about one fifth. The reactions studied include those of copper, phosphorus and iron with air.
C2d.3	**The preparation and properties of oxygen** Students make oxygen from the catalytic decomposition of hydrogen peroxide. You may use the effect of the oxygen collected on a glowing and lighted splint to illustrate the importance of oxygen in supporting combustion.
C2d.4	**The preparation and properties of carbon dioxide** This activity involves preparing carbon dioxide by two different methods: the action of dilute hydrochloric acid on calcium carbonate and the decomposition of copper(II) carbonate. Lime water is introduced as a means of identifying the carbon dioxide. The properties of carbon dioxide are then linked to its uses.

C2d.5	**Carbon dioxide as a greenhouse gas** This activity focuses on carbon dioxide as a greenhouse gas and its potential contribution to climatic change. It provides an opportunity to research recent contributions to the debate on climatic change.
C2d.6	**Consolidation and summary** This activity provides an opportunity for a recap on the learning points encountered in the topic and for the students to answer the End of Topic Questions in the Student Book.

Activity C2d.1 Orientation

Learning objectives

- To revise the concept of basic and acidic oxides
- To identify the gases which are the main components of the air

Learning outcomes

- Know that a basic oxide is the oxide of a metal and an acidic oxide is the oxide of a non-metal
- Know that the air contains approximately 78% nitrogen, 21% oxygen, 0.9% argon and 0.04% carbon dioxide
- Be able to describe the reactions of oxygen with magnesium, carbon and sulfur and write equations for these reactions

Common misconceptions

Some students think that air contains a much higher proportion of carbon dioxide than 0.04%.

Resources

Student Book pages 134–135

Files on CD–ROM: C2d.1a_tech_notes

Resources for a demonstration (see below)

Approach

Ask students where oxygen is found in the Periodic Table. What does its position say about its properties? Look at the chart on page 134 in the Student Book showing the components of air.

Emphasise that oxygen is a very special gas because it supports combustion – without it, materials couldn't burn in the air as they do. Explain that oxygen is highly reactive and forms oxides with most elements, except some of the least reactive metals (such as gold) and the noble gases.

Demonstrate the reaction of oxygen with magnesium, phosphorus, sulfur and carbon, heating the element on a combustion spoon before putting into a gas jar of oxygen. (Use a fume cupboard or a very well ventilated part of the laboratory). Test the products with damp universal indicator paper. Ask students what the oxides of metals and non-metals are called (basic and acidic oxides). Ask students to write chemical equations for the reactions demonstrated.

Technician's notes

You will need the following resources for a demonstration.

Be sure to check the latest safety notes on these resources before proceeding.

magnesium ribbon
phosphorus (white)
sulfur
carbon powder
4 gas jars of oxygen
combustion spoons
Bunsen burner and heat resistant mat
eye protection and safety screen

Activity C2d.2 Determining the percentage composition of oxygen in the air

Learning objectives

- To demonstrate an experiment to determine the percentage of oxygen in the air
- To consider other methods of determining the percentage of oxygen in the air

Learning outcomes

- Be able to explain how the reactions of copper, iron and phosphorus with air can be used to determine the percentage by volume of oxygen in the air
- Know that oxygen accounts for approximately one-fifth of the air

Resources

Student Book pages 135–136, 138

Files on CD–ROM: C2d.3a_tech_notes

Resources for a demonstration (see below)

Approach

Demonstrate the experiment to determine the percentage of oxygen in the air by repeatedly passing air over heated copper turnings. **Emphasise the importance of letting the apparatus cool down before taking the final volume of remaining air.**

Look at the diagram on page 135 in the Student Book showing the classic experiment using phosphorus. The *Developing Investigative Skills* feature on page 138 relates to a similar experiment with iron – students could answer the associated questions now, or alternatively you might like to leave this until the consolidation and summary activity.

Technician's notes

You will need the following resources for a demonstration.

Be sure to check the latest safety notes on these resources before proceeding.

copper turnings in a silica tube
2 gas syringes connected as on page 135 in the Student Book
Bunsen burner
eye protection

Answers

Page 136

1. Argon
2. 0.04%
3. Nitrogen
4. Copper(II) oxide

Developing investigative skills, page 138

1.

Volume readings	Volume of gas (cm^3)
Initial volume of air	100
Final volume of air	85
Reduction in volume	15

2. The iron filings would have turned orange as the iron rusted.

3. The iron has reacted with oxygen (it has rusted) and formed iron(III) oxide.

4. $4Fe(s) + 3O_2(g) \rightarrow 2Fe_2O_3(s)$

5. a) The reaction may not be complete – some of the iron had not reacted.

   b) The air might not have been a normal sample (for example, it might have been expired air which contains less oxygen).

   c) Experimental error in setting up the experiment and taking the readings.

6. To equalise the pressure so both readings were taken at the same pressure (the volume of a gas will change as the pressure changes).

Activity C2d.3 The preparation and properties of oxygen

Learning objectives

- To prepare oxygen and identify its presence using a glowing splint

Learning outcomes

- Be able to describe the laboratory preparation of oxygen from hydrogen peroxide using manganese(IV) oxide as a catalyst
- Know that the relighting of a glowing splint is used as a simple identification test for oxygen

Resources

Student Book page 136–137

Worksheet C2d.3a The preparation and properties of oxygen

Files on CD–ROM: C2d.3a_worksheet, C2d.3a_tech_notes

Resources for a class practical (see below)

Approach

In groups, students can prepare oxygen by adding hydrogen peroxide to manganese(IV) oxide. The oxygen can be collected by displacement of water. Explain that the manganese(IV) oxide is a catalyst and speeds up the decomposition of the hydrogen peroxide.

CAUTION: manganese(IV) oxide is harmful. You and your students must take care to avoid inhaling when adding the fine powder to the conical flask.

Students should be able to collect several test tubes of gas and test different test tubes of the gas with a glowing splint and then a lighted splint. Explain that the use of a glowing splint is the recognised identification test for oxygen.

Worksheet C2d.3a can be used to support this activity. Alternatively, it could be used in the Consolidation and summary activity as a way of revising the ideas on preparation, properties and uses of oxygen and carbon dioxide.

Technician's notes

You will need the following resources for a class practical.

Be sure to check the latest safety notes on these resources before proceeding.

20 volume hydrogen peroxide
manganese(IV) oxide
conical flask, tap funnel or thistle funnel, delivery tube and bung, trough and test tubes
spatula
eye protection

Answers

Page 137

1. a) A catalyst is a substance that increases the rate of a reaction.

   b) Manganese (IV) oxide.

2. A basic oxide will react with an acid to form a salt.

3. $2Ca(s) + O_2(g) \rightarrow 2CaO(s)$

4. Acidic oxide. Phosphorus is a non-metal.

C2d.3a The preparation and properties of oxygen

This worksheet focuses on the laboratory preparation of oxygen. It also asks you to describe the test for oxygen. The final section is concerned with the uses of oxygen and the properties that make it suited to these uses.

Apparatus

20 volume hydrogen peroxide

manganese(IV) oxide

conical flask, tap funnel or thistle funnel, delivery tube and bung, trough and test tubes

spatula

eye protection

SAFETY INFORMATION
Manganese(IV) oxide is harmful.
Wear eye protection.

Demonstrate and describe techniques

Oxygen can be collected by displacement of water, as shown in the diagram.

1. Put on your eye protection.

2. Set up the apparatus as shown in the diagram.

3. Put the manganese(IV) oxide into the conical flask. Then add the hydrochloric acid to the thistle funnel.

4. Collect the gas in the test tube.

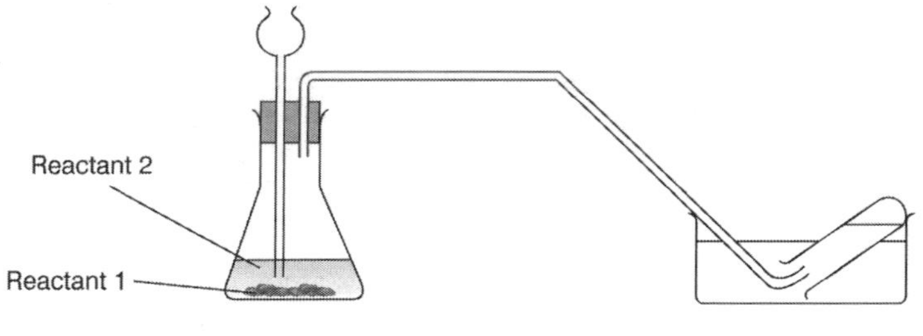

Page 1 of 2

Make observations and measurements

The method above can be used to prepare oxygen. The gases can then be tested to confirm their presence. Complete the table below for the preparation and testing oxygen.

Gas	Preparation: reactants needed		Test for gas	
	Reactant 1	Reactant 2	Description of test	Result of test
Oxygen				

Analyse and interpret data

1. A catalyst is used in the preparation of oxygen. What is a catalyst?

2. Oxygen can be collected by displacement of water. What property of the gases makes this a feasible method of collection?

3. Use some research to find out about the uses of oxygen. Complete the table below to show the uses of oxygen and the properties that makes it suited to that use.

Gas	Uses of the gas	Properties of gas suited to use
Oxygen	1. 2. 3.	1. 2. 3.

Activity C2d.4 The preparation and properties of carbon dioxide

Learning objectives

- To prepare and identify carbon dioxide
- To relate the uses of carbon dioxide to its properties

Learning outcomes

- Be able to describe the preparation of carbon dioxide from calcium carbonate and dilute hydrochloric acid
- Be able to describe the formation of carbon dioxide from the thermal decomposition of copper(II) carbonate
- Be able to explain that carbon dioxide is used as the gas in fizzy drinks as it dissolves in water
- Be able to explain that carbon dioxide is used in fire extinguishers as it is denser than air

Resources

Student Book pages 138–140

Worksheet C2d.4a The preparation and properties of carbon dioxide

Files on CD–ROM: C2d.4a_worksheet, C2d.4a_tech_notes

Resources for a class practical (see below)

Approach

Students can prepare and test carbon dioxide using two different reactions:

1. Marble chips and dilute hydrochloric acid

2. Heating copper(II) carbonate

Worksheet C2d.4a can be used to support this activity. Alternatively, it could be used in the Consolidation and summary activity as a way of revising the ideas on preparation, properties and uses of oxygen and carbon dioxide.

Ask students to write down the equations for the two reactions used to prepare the carbon dioxide. Explain the reactions that occur when lime water is used to test for carbon dioxide. If you are using Worksheet C2d.4a at this stage, it includes a focus on the uses of carbon dioxide and the properties of the gas that make it suitable for these uses.

Technician's notes

You will need the following resources for a class practical.

Be sure to check the latest safety notes on these resources before proceeding.

dilute hydrochloric acid (1M)
calcium carbonate/marble chips
copper(II) carbonate
lime water
conical flask, tap funnel or thistle funnel, delivery tube and bung, trough and test tubes
test tube and delivery tube, test tube holder
Bunsen burner and heatproof mat
eye protection

C2d.4a The preparation and properties of carbon dioxide

This worksheet focuses on the laboratory preparation of carbon dioxide. It also asks you to describe the test for carbon dioxide. The final section is concerned with the uses of carbon dioxide and the properties that it suited to these uses.

Apparatus

dilute hydrochloric acid

calcium carbonate/marble chips

copper(II) carbonate

lime water

conical flask, tap funnel or thistle funnel, delivery tube and bung, trough and test tubes

test tube and delivery tube, test tube holder

Bunsen burner and heat proof mat

eye protection

SAFETY INFORMATION
Copper(II) carbonate is harmful.
Wear eye protection.

Demonstrate and describe techniques

Carbon dioxide can be collected by displacement of water.

Method 1

This method is suitable for the preparation of both oxygen and carbon dioxide.

1. Put on your eye protection.
2. Set up the apparatus as shown in the diagram.
3. Put the marble chips into the conical flask. Then add the hydrochloric acid to the thistle funnel.
4. Collect the gas in the test tube.

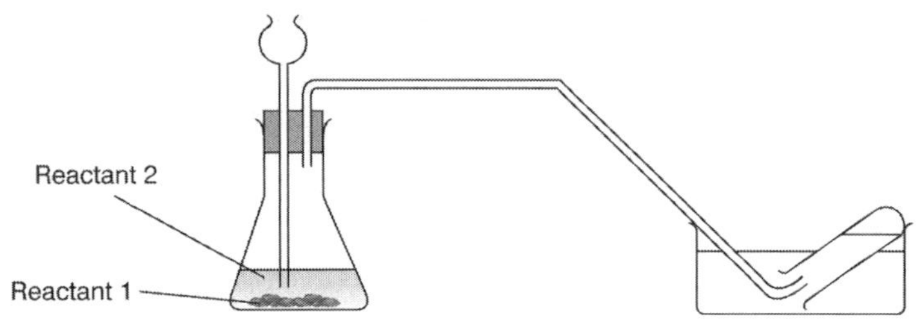

Page 1 of 2

Method 2

This method is only suitable for the preparation of carbon dioxide.

1. Set up the apparatus as shown in the diagram. Put copper carbonate in the test tube.
2. Heat the copper carbonate and collect the gas in a test tube as shown.
3. Remove the delivery tube from the water **before** you stop heating.

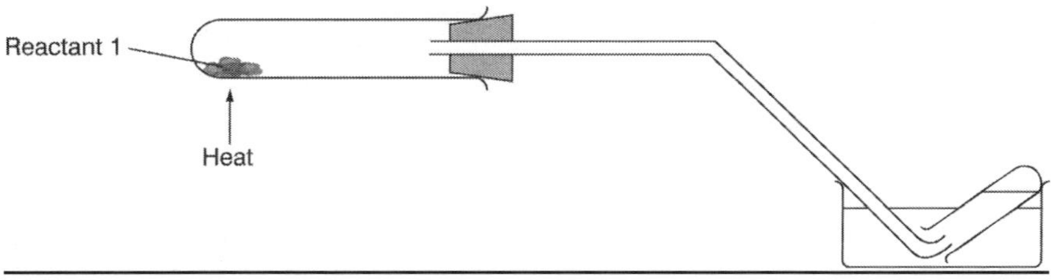

Make observations and measurements

The methods above can be used to prepare carbon dioxide. The gas can then be tested to confirm that it is carbon dioxide. Complete the table below for the preparation and testing of carbon dioxide.

Gas	**Preparation: reactants needed**		**Test for gas**	
	Reactant 1	**Reactant 2**	**Description of test**	**Result of test**
Carbon dioxide (Method 1)				
Carbon dioxide (Method 2)				

Analyse and interpret data

1. Carbon dioxide can be collected by displacement of water. What property of the gas makes this a feasible method of collection?

2. Do some research to find out about the uses of carbon dioxide. Complete the table below to show the uses and the properties of the gas that makes it suited to that use.

Gas	**Uses of the gas**	**Properties of gas suited to use**
Carbon dioxide	1. 2. 3.	1. 2. 3.

Activity C2d.5 Carbon dioxide as a greenhouse gas

Learning objectives

- To consider the potential detrimental effects of the formation of carbon dioxide from the combustion of fuels.

Learning outcomes

- Know that carbon dioxide is a greenhouse gas and may contribute to climate change.

Resources

Student Book pages 140–142

Approach

Students will be familiar with the production of carbon dioxide from the burning of carbon. Discuss the hypothesis that as a greenhouse gas carbon dioxide may be responsible for global warming. Explain that scientists are split on this issue and some think that observed rises in average temperatures are just part of the normal cycle of climatic change and are not associated with greenhouse gases.

Additional information is provided in the Student Book on page 140. Students could usefully undertake some research on carbon dioxide as a greenhouse gas and other gases that behave in the same way.

Answers

Page 141

1. Bubble the gas through lime water. A white precipitate indicates the presence of carbon dioxide.
2. Dry ice is solid carbon dioxide.
3. Carbonic acid.

Page 142

1. a) $2H_2O_2(aq) \rightarrow 2H_2O(l) + O_2(g)$

   b) Manganese(IV) oxide. A catalyst will speed up the rate of a chemical reaction. It does this by providing an alternative route for the chemical reaction which has a lower activation energy.

   c) Description to include:

      Suitable apparatus to collect oxygen (ref. Fig. 2.27)

      Pour hydrogen peroxide into conical flask

      Cut small strips of liver to act as catalyst: weigh and record mass

      Add catalyst to hydrogen peroxide: replace stopper and collect oxygen until reaction stops

      Remove liver from solution and reweigh/or repeat a further experiment using the same pieces of liver

      Mass of liver should be (the same)/liver can be re-used

   d) Oxygen can be collected over water as it is only slightly soluble. Oxygen will rise to the top of a gas jar filled with water and displace the water by 'downward displacement'.

   e) Oxygen can be obtained by the fractional distillation of liquid air.

2. a) When magnesium burns it reacts with oxygen to form magnesium oxide. Pure oxygen is 100% oxygen gas; the air is only 21% oxygen.

b) Magnesium nitride Mg_3N_2

c) Magnesium oxide is a base. Bases will react with acids to form salt + water. If dilute hydrochloric acid is added to magnesium oxide (white powder) it will dissolve to form a colourless solution of magnesium chloride (a soluble salt) + water.

3. a) carbon dioxide can be prepared by

   Adding dilute acid to a metal carbonate/metallic carbonate

   Decomposition of a metal carbonate/metallic carbonate

b) Description to include:

   Suitable apparatus (see Fig. 2.31)

   Use sodium carbonate and dilute hydrochloric acid.

   Collect carbon dioxide in gas jars over water.

c) $Na_2CO_3(s) + 2HCl(aq) \rightarrow$ _2NaCl(aq) + $H_2O(l) + CO_2(g)$

Activity C2d.6 Consolidation and summary

Learning objectives

- To review the learning points of the topic
- To test understanding through answering questions

Learning outcomes

- Be familiar with the knowledge and understanding summarised in the End of Topic Checklist
- Be able to apply this knowledge and understanding by answering End of Topic Questions in the Student Book

Resources

Student Book pages 143–144

Worksheet C2d.3a

Worksheet C2d.4a

Files on CD–ROM: C2d.3a_worksheet, C2d.4a_worksheet

Approach

You could use the *Developing Investigative Skills* feature on page 138 of the Student Book to revise ideas on the composition of the air if it was not used in Activity C2d.2. Similarly, Worksheets C2d.3a and C2d.4a could be used if they have not been used to support Activities C2d.3 or C2d.4.

It is worth spending some time revising some of the equations encountered during the topic. Focus on the End of Topic Checklist and End of Topic Questions in the Student Book as a way of recapping on the main learning points.

Answers

End of Topic Questions mark scheme

Question	Correct answer	Marks
1 a)	21% (allow 1/5)	1 mark
1 b)	0.04%	1 mark
2 a)	When air is passed over heated copper the oxygen in the air reacts with the copper to form copper(II) oxide. As the oxygen is removed from the air the volume of gas is reduced eventually by about 1/5).	1 mark 1 mark
2 b)	In the presence of moisture, oxygen in the air reacts with iron to form iron(III) oxide. As the oxygen is removed from the air, external air pressure causes the water to rise up the tube.	1 mark 1 mark
3	Manganese(IV) oxide acts as a catalyst. It increases the rate at which the hydrogen peroxide decomposes to form oxygen.	1 mark 1 mark
4 a)	i) $4Na(s) + O_2(g) \rightarrow 2Na_2O(s)$ (1 mark for formulae; 1 mark for balancing) ii) $4Al(s) + 3O_2(g) \rightarrow 2Al_2O_3(s)$ (1 mark for formulae; 1 mark for balancing) iii) $2Mg(s) + O_2(g) \rightarrow 2MgO(s)$ (1 mark for formulae; 1 mark for balancing)	2 marks 2 marks 2 marks
4 b)	The oxides are basic – they are oxides of metals.	1 mark
5	Calcium carbonate (allow other metal carbonates). Dilute hydrochloric acid (allow dilute nitric acid and dilute sulfuric acid providing the salt formed is not insoluble: for example, don't allow sulfuric acid with calcium carbonate.)	1 mark 1 mark
6 a)	Decomposition means the breaking down of a compound to form smaller compounds).	1 mark
6 b)	$CuCO_3(s) \rightarrow CuO(s) + CO_2(g)$ (1 mark for formulae; 1 mark for balancing)	2 marks
7	Carbon dioxide does not support combustion (do not allow 'it does not burn'). It is denser than air and so settles on a fire and smothers it (excludes the air).	1 mark 1 mark
8	A greenhouse gas prevents some of the energy on the Earth's surface being lost from the Earth's atmosphere. (It absorbs and emits radiation.) Such gases are thought to be responsible for global warming.	1 mark 1 mark
	Total:	24 marks

C2e Hydrogen and water

Introduction

Hydrogen does not belong to a group of elements and is often placed on its own the Periodic Table. Its electronic structure has some similarities with the Group 1 elements, but hydrogen is not a metal, although it usually forms positive ions.

The topic focuses on preparing hydrogen from the metal and acid reaction and then its reaction with oxygen to form water. The test for the presence of water is introduced, together with a means of identifying pure water. The method of preparation has obvious links to Topic 2f *Reactivity series* and also to Topic 1i *Electrolysis*, as hydrogen is often discharged at the cathode in the electrolysis of aqueous solutions.

This is a very short topic with some limited opportunities for developing investigative skills.

Links to other topics

Sections	Essential background knowledge	Useful links
1 Principles of chemistry	1e Chemical formulae and chemical equations	1i Electrolysis
2 Chemistry of the elements	2a The Periodic Table	2f Reactivity series 2g Tests for ions and gases

Topic overview

C2e.1	**Orientation** The purpose of this very short activity is to highlight the unusual positioning of hydrogen in the Periodic Table. The unique atomic structure of the hydrogen atom is revised (one of the isotopes contains no neutrons in the nucleus) and the significance of its single electron.
C2e.2	**The preparation of hydrogen** This activity concentrates on preparing hydrogen from a metal and dilute hydrochloric or sulfuric acid. It provides an opportunity to compare the relative reactivity of a series of metals and to introduce the test used to confirm the identity of hydrogen. Emphasis is also placed on writing chemical equations for the reactions encountered.
C2e.3	**Water** This activity demonstrates the formation of water when hydrogen burns and then introduces the use of anhydrous copper(II) sulfate to detect the presence of water. The measurement of the boiling point of a liquid is introduced as a means of identifying pure water.
C2e.4	**Consolidation and summary** This activity provides an opportunity for a quick recap on the ideas encountered in the topic and time for the students to answer the End of Topic Questions in the Student Book.

Activity C2e.1 Orientation

Learning objectives

- To identify the position of hydrogen in the Periodic Table
- To revise the atomic structure of hydrogen

Learning outcomes

- Know that hydrogen is not placed in a group in the Periodic Table
- Know the atomic structure of hydrogen

Common misconceptions

Some students think that hydrogen atoms do not contain neutrons. (Only one isotope of hydrogen does not have any neutrons.)

Resources

Student Book page 145

Approach

This introductory activity is very short and should only need a few minutes of teaching time. Ask students to write down as much as they know about hydrogen, including its atomic structure. Identify the position of hydrogen in the Periodic Table and the fact that it is often positioned on its own, not as part of a group of elements. Explain that hydrogen gas exists as diatomic molecules (H_2).

Activity C2e.2 The preparation of hydrogen

Learning objectives

- To prepare hydrogen and identify its presence with a lighted splint
- To investigate the different reactivities of metals with dilute acids

Learning outcomes

- Be able to describe how hydrochloric acid reacts with magnesium, aluminium, zinc and iron to produce hydrogen and a metal chloride
- Be able to describe how sulfuric acid reacts with magnesium, aluminium, zinc and iron to produce hydrogen and a metal sulfate
- Be able to describe a simple test to identify hydrogen

Resources

Student Book page 145–147

Worksheet C2e.2a The preparation of hydrogen

Files on CD–ROM: C2e.2a_worksheet, C2e.2a_tech_notes

Resources for a class practical and demonstration (see below)

Approach

This activity has two parts.

Part 1

Initially students can use Worksheet C2e.2a to investigate the reactions of dilute hydrochloric and sulfuric acids with a series of metals. They should be able to identify a rough order of reactivity but this may not be absolutely correct, given that the surface areas of the metals will not be the same.

Encourage students to write the equations suggested on the worksheet and also look at those in the Student Book on page 146. Emphasise that hydrochloric acid forms the metal chloride, whereas sulfuric acid forms the metal sulfate.

Part 2

Demonstrate the preparation of hydrogen using dilute hydrochloric acid and granulated zinc, as shown in the Student Book on page 146. Explain the function of the copper (or copper(II) sulfate). Emphasise the importance of keeping a flame well away from the apparatus.

Explain that pure hydrogen will burn (it can be burned at a jet), but when hydrogen is mixed with air a flame can produce an explosion. Refer to the Hindenburg disaster in the Student Book on page 148.

Technician's notes

You will need the following resources for a class practical and demonstration.

Be sure to check the latest safety notes on these resources before proceeding.

Part 1
test tubes, test tube rack and droppers
small strips of magnesium ribbon
small strips of aluminium, copper and lead foil
granulated zinc

iron powder
0.5M hydrochloric acid
0.5M sulfuric acid
Part 2
apparatus for preparation of hydrogen (see Student Book page 146)
1M copper(II) sulfate solution
granulated zinc
1M hydrochloric acid
eye protection

Hazcard note:

For large-scale reduction of copper (II) oxide:

- Wear a face shield.
- Students should wear eye protection and be kept at least 2 m from the apparatus.
- Use safety screens.
- Do not use concentrated sulfuric(VI) acid as a drying agent.

This demonstration has resulted in accidents caused by hydrogen/air explosions, one of which resulted in the prosecution of a teacher. Use anhydrous calcium chloride, loosely packed in a U-tube for the drying agent. Ensure the apparatus is flushed free of air with hydrogen before lighting Bunsen burner. Practice the procedure before demonstrating it in front of a class.

Answers

Page 147

1. aluminium + hydrochloric acid → aluminium chloride + hydrogen
2. $2Al(s) + 6HCl(aq) \rightarrow 2AlCl_3(aq) + 3H_2(g)$
3. Hydrochloric acid forms salts called chlorides.

C2e.2a The preparation of hydrogen

Hydrogen can be prepared by the reaction between a metal and dilute hydrochloric acid or dilute sulfuric acid. In this activity you will be able to compare the reactivity of some different metals with these two dilute acids.

The quantities of hydrogen produced will be relatively small and so will not pose a significant hazard.

Apparatus

small samples of: aluminium, copper, iron, lead, magnesium and zinc

dilute hydrochloric acid, dilute sulfuric acid

test tubes and test tube rack, wooden splint

eye protection

SAFETY INFORMATION

Magnesium is highly flammable.

Wear eye protection.

Method

1. Put on your eye protection.
2. Put a sample of each metal in a separate test tube and add about a 2 cm depth of dilute hydrochloric acid. Record your observations in the table.
3. Repeat the procedure using the metals and dilute sulfuric acid.
4. Where the reaction is sufficiently vigorous, collect some gas by putting a bung in the end of the test tube for 30 seconds and then remove it and apply a lighted splint to the mouth of the test tube.

Make observations and measurements

Metal	Observations	
	Dilute hydrochloric acid	**Dilute sulfuric acid**
Aluminium		
Copper		
Iron		
Lead		
Magnesium		
Zinc		

Page 1 of 2

Analyse and interpret data

4. Use your observations to put the metals in an order of reactivity:

Most reactive ______________

____

____

____

____

Least reactive ______________

5. Complete the table below:

Test for hydrogen gas	
What you use for the test	The results that confirms the presence of hydrogen

6. Write equations for the following reactions:

a) Zinc with dilute hydrochloric acid

____

b) Magnesium with dilute sulfuric acid

____

Evaluate data and methods

7. The method you have used may not produce the correct order of reactivity for the metals you tested. Suggest how your method could be improved to give greater accuracy.

____

____

____

____

Page 2 of 2

Activity C2e.3 Water

Learning objectives

- To demonstrate the formation of water when hydrogen burns
- To know how to identify the presence of water and verify that the water is pure

Learning outcomes

- Be able to describe how hydrogen burns in air or oxygen to form water
- Be able to describe how anhydrous copper(II) sulfate can be used to identify the presence of water and that the anhydrous copper(II) sulfate changes colour from white to blue when water is present
- Be able to describe a test to show whether water is pure or not

Resources

Student Book pages 146–148

Files on CD–ROM: C23.3a_tech_notes

Resources for a demonstration (see below)

Approach

If a hydrogen cylinder is available, demonstrate the formation of water when hydrogen is burned – if the flame is positioned under a cold surface, drops of water (condensation) form. Refer to the equation for the reaction.

If you use a hydrogen cylinder, it is ESSENTIAL that all air is 'flushed out' of the apparatus/delivery tube before you attempt to light the hydrogen at a jet. It is NOT safe to try and burn hydrogen that is generated by a reaction in the laboratory – it is practically impossible to 'flush out' all the air from the apparatus.

Demonstrate the action of heat on hydrated copper(II) sulfate crystals to produce anhydrous copper(II) sulfate. Stop heating as soon as the colour change to white is complete/stops. Illustrate the reversibility of the reaction by adding water to the anhydrous salt. Explain the terms 'anhydrous', 'hydrated' and 'water of crystallisation'. Emphasise that anhydrous copper(II) sulfate will detect the presence of water but will not indicate whether a liquid is pure water. Stop heating when the colour change stops.

Finally, demonstrate the technique for determining the boiling point of a liquid using apparatus similar to that shown in the Student Book on page 147. Explain that impurities will increase the boiling point of pure water (100 °C at standard pressure).

Hazcard note:

For large-scale reduction of copper (II) oxide:

- Wear a face shield.
- Students should wear eye protection and be kept at least 2 m from the apparatus.
- Use safety screens.
- Do not use concentrated sulfuric(VI) acid as a drying agent.

This demonstration has resulted accidents, caused by hydrogen/air explosions, one of which resulted in the prosecution of a teacher. Use anhydrous calcium chloride, loosely packed in a U-tube for the drying agent. Ensure the apparatus is flushed free of air with hydrogen before lighting Bunsen burner. Practice the procedure before demonstrating it in front of a class.

Technician's notes

You will need the following resources for demonstration purposes.

Be sure to check the latest safety notes on these resources before proceeding.

hydrogen cylinder
delivery tube, jet, cold surface (flask full of cold water)
safety screen
copper(II) sulfate crystals
test tube and holder
Bunsen burner
boiling point apparatus (see Student Book page 147)
eye protection

Answers

Page 148

1. Without water/without water of crystallisation
2. Determine the boiling point of the liquid. (Pure water boils at 100 ºC at normal pressure).
3. $2H_2(g) + O_2(g) \rightarrow 2H_2O(l)$

Activity C2e.4 Consolidation and summary

Learning objectives

- To review the learning points of the topic
- To test understanding through answering questions

Learning outcomes

- Be familiar with the knowledge and understanding summarised in the End of Topic Checklist
- Be able to apply this knowledge and understanding by answering the End of Topic Questions in the Student Book

Resources

Student Book page 149–150

Approach

Further work on the reactivity series is covered in Topic 2f *Reactivity Series*. Focus on the End of Topic Questions to test understanding of the requirements of the Checklist.

Answers

End of Topic Questions mark scheme

Question	Correct answer	Marks
1 a)	Effervescence/bubbling.	1 mark
	The magnesium slowly disappears/gets smaller.	1 mark
1 b)	i) $Mg(s) + H_2SO_4(aq) \rightarrow MgSO_4(aq) + H_2(g)$	2 marks
	(1 mark for formulae; 1 mark for balancing)	
	ii) $Fe(s) + H_2SO_4(aq) \rightarrow FeSO_4(aq) + H_2(g)$	2 marks
	(1 mark for formulae; 1 mark for balancing)	
	iii) $2Al(s) + 6HCl(aq) \rightarrow 2AlCl_3(aq) + 3H_2(g)$	2 marks
	(1 mark for formulae; 1 mark for balancing)	
2	Put a lighted splint into the gas.	1 mark
	The gas burns with a pop.	1 mark
3	Add anhydrous copper(II) sulfate to the liquid.	1 mark
	The anhydrous copper(II) sulfate turns from white (very pale blue) to deep blue.	1 mark
4	Heat a sample of water in a test tube to its boiling point.	1 mark
	If the water is pure, a thermometer in the boiling water will record 100 °C (at normal pressure).	1 mark
5 a)	$2H_2(g) + O_2(g) \rightarrow 2H_2O(l)$	2 marks
	(1 mark for formulae; 1 mark for balancing)	
5 b)	Advantage: the product of burning the fuel is water (environmentally friendly).	1 mark
	Disadvantage: the gas is highly flammable and extreme care needs to be taken during storage and filling of the car.	1 mark
	Total:	18 marks

C2f Reactivity series

Introduction

Students will be aware of the differences in the reactivity of metals from their work on the Periodic Table and the Group 1 elements. The will also have encountered the use of the metal and acid reaction to produce hydrogen.

This topic introduces competition or displacement reactions in the context of metals and metal oxides, as well as metals and solutions of metal ions. You will introduce redox reactions and the definitions of oxidation and reduction in terms of the gain or loss of oxygen.

Finally, the topic covers the rusting of iron and steel and methods for limiting or preventing rusting, including the use of galvanising, which depends on the greater reactivity of zinc compared to iron. There are links to Topic 1i *Electrolysis*, Topic 4a *Acids, alkalis and salts* and Topic 5a *Extraction and uses of metals*. There are some opportunities for students to develop their investigative skills.

Links to other topics

Sections	**Essential background knowledge**	**Useful links**
1 Principles of chemistry		1i Electrolysis
2 Chemistry of the elements	2a The Periodic Table 2b Group 1 elements 2e Hydrogen and water	
4 Physical chemistry		4a Acids, alkalis and salts
5 Chemistry in society		5a Extraction and uses of metals

Topic overview

C2f.1	**Orientation** This activity provides an opportunity to revise the reactions of the Group 1 metals with water and the use of metals and dilute hydrochloric or sulfuric acids to make hydrogen. You may demonstrate the reaction of sodium and water and refer to previous work on metals and acids. Also demonstrate the reaction of magnesium and steam to illustrate the relatively high reactivity of magnesium.
C2f.2	**Displacement reactions** This activity introduces displacement reactions involving a metal and metal oxide through demonstrating the reaction between magnesium and copper(II) oxide. Use reactions of this type to introduce or revise the concept of a redox reaction. Next, introduce displacement reactions in solution involving a metal and a solution of a metal salt. Worksheet C2f.2 reinforces the concept of a displacement reaction and provides an opportunity to use chemical equations to summarise the outcomes of these reactions.
C2f.3	**Limiting corrosion** This activity focuses on the problems caused by the rusting of iron and steel and the relative merits of the methods used to prevent or limit such corrosion. Worksheet C2f.3 provides students with a relatively straightforward plan and design activity.

C2f.4	**Consolidation and summary** This activity provides an opportunity for a quick recap on the ideas encountered in the topic and time for the students to answer the End of Topic Questions in the Student Book.

Activity C2f.1 Orientation

Learning objectives

- To recap on existing knowledge of the differing reactivities of metals with water and dilute hydrochloric or sulfuric acids

Learning outcomes

- Understand that metals can be arranged in a reactivity series: potassium, sodium, lithium, calcium, magnesium, aluminium, zinc, iron, copper, silver and gold
- Be able to describe how the reactions of metals with water and dilute acids can be used to deduce an order of reactivity

Resources

Student Book pages 151–152, 156

Files on CD–ROM: 2f.1a_tech_notes

Resources for a demonstration (see below)

Approach

If appropriate, you may demonstrate the reaction of sodium with water to revise the highly reactive nature of the Group 1 metals. Similarly, demonstrate the reaction of magnesium with steam to illustrate magnesium's relatively high position in the reactivity series.

The reaction of magnesium with steam can most easily be demonstrated using a Pyrex® test tube with a plug of water soaked mineral wool in the bottom, a coil of magnesium ribbon in the middle and with a bung and small jet positioned in the open end of the tube.

The tube is clamped horizontally and the magnesium is heated strongly with the flame occasionally being directed for a few seconds to the mineral wool. When the reaction starts hydrogen can be ignited at the jet.

Remind students of the reactions of metals with dilute hydrochloric or sulfuric acids to produce hydrogen. Refer to the reactivity series on page 152 of the Student Book. Point out that the metals at the bottom of the series are very unreactive and are found in nature in their native states. Refer to the *Science in Context* feature on page 156 of the Student Book.

Technician's notes

You will need the following resources for a demonstration.

Be sure to check the latest safety notes on these resources before proceeding.

sodium
large trough of water, safety screen
magnesium ribbon
mineral wool, test tube, bung and jet
eye protection

Activity C2f.2 Displacement reactions

Learning objectives

- To introduce or reinforce ideas on competition or displacement reactions
- To introduce the characteristics of a redox reaction

Learning outcomes

- Understand how displacement reactions between metals and metal oxides and between metals and metal salts in aqueous solution can be used to deduce the order of reactivity of metals
- Understand that oxidation is the addition of oxygen
- Understand that reduction is the removal of oxygen
- Understand that an oxidising agent will oxidise another element or compound but will itself be reduced
- Understand that a reducing agent will reduce another element or compound but will itself be oxidised
- Understand the term redox

Resources

Student Book page 152–154

Worksheet C2f.2a Metals and displacement reactions

Files on CD–ROM: C2f.2a_worksheet, C2f.2a_tech_notes

Resources for a demonstration and a class practical (see below)

Approach

Demonstrate the reaction of magnesium powder with copper(II) oxide (in a fume cupboard: the reaction is potentially quite violent) as an example of a displacement or competition reaction. Use the chemical equation to introduce or revise the concepts of a redox reaction, oxidation, reduction, oxidising agent and reducing agent.

Use Worksheet C2f.2a to support student practical work on displacement reactions involving a metal and a solution containing the ions of another metal. As part of the activity, ask students to use the reactivity series to make predictions about displacement reactions. The *Developing Investigative Skills* feature on page 154 also focuses on displacement reactions in solution. You could leave this until the consolidation and summary activity, where you could use it as a revision exercise.

Technician's notes

You will need the following resources for a demonstration and a class practical.

Be sure to check the latest safety notes on these resources before proceeding.

Demonstration
magnesium powder and spatula
copper(II) oxide and spatula
crucible, pipe clay triangle, tripod and Bunsen burner
fume cupboard or safety screen
eye protection

Student practical
small samples of magnesium ribbon
small samples of copper, iron, lead and zinc
0.1M lead(II) nitrate solution
0.1M copper(II) sulfate
0.1M iron(II) ammonium sulfate labelled iron(II) sulfate
0.1 M magnesium sulfate and 0.1M zinc sulfate
test tubes and test tube racks
eye protection

Note:

For the **reduction of the oxide(II) using magnesium**:

- Wear goggles or a face shield.
- Use safety screens.
- Do not go back to a seemingly failed reaction.

There can be a considerable delay before the violent reaction takes place. If the reaction fails to ignite, turn the gas off using the emergency switch or isolation valve.

For the **reaction of the metal powder with copper oxide**:

- Wear goggle or a face shield.
- Use safety screens.
- The room should be well ventilated to remove the smoke.
- Protect the bench with hard-board or heat-proof mats.
- Students should stand towards the back of the room.
- This is best preformed at the end of a lesson after which the class leaves.

Instructions:

- Weigh 0.5 g of each reagent on paper.
- Mix by passing them from one piece of paper to the other ad place the mixture in a bottle top (the plastic insert should be burnt out first).
- Place the bottle top on a pipe-clay triangle supported by a tripod.
- Keep heating strongly with a Bunsen burner, retire to where students are standing and watch carefully; the induction period may be a few seconds or as long as a minute.

If the reaction fails to work (probably the magnesium powder has oxidised). DO NOT RETURN TO THE APPARATUS AND POKE THE MIXTURE. It is best to turn the gas off at the main control for the laboratory and allow the mixture to cool before adding to water. Place used bottle tops in the waste.

Answers

Page 153

1. No. Copper is below hydrogen in the reactivity series.
2. $2K(s) + 2H_2O(l) \rightarrow 2KOH(aq) + H_2(g)$
3. $Mg(s) + PbO(s) \rightarrow MgO(s) + Pb(s)$
4. No. Carbon is below magnesium in the reactivity series.

Developing investigative skills, page 154

1. A metal cannot displace itself from a solution of its salt.
2. She already knows that metal D cannot displace the other three metals and so must be the least reactive metal.
3. B, C, A, D
4. 10 – yes, 11 – yes, 12 – yes
5. $B(s) + C(NO_3)_2(aq) \rightarrow B(NO_3)_2(aq) + C(s)$
6. Copper.

C2f.2a Metals and displacement reactions

A more reactive metal will displace a less reactive metal from a solution containing ions of the less reactive metal. These reactions have been used to confirm the order of metals in the reactivity series.

In this activity you should be able to use your knowledge of the reactivity series to make predictions about the likelihood of a displacement reaction taking place.

Apparatus

small samples of the following metals: copper, iron, lead, magnesium and zinc

solutions of copper(II) sulfate, iron(II) sulfate, lead(II) nitrate, magnesium sulfate, zinc sulfate

test tubes and test tube rack

eye protection

SAFETY INFORMATION

Magnesium is highly flammable.

Lead(II) nitrate is toxic.

Copper(II) sulfate and iron(II) sulfate are harmful.

Wear eye protection.

Method

1. Put on your eye protection.
2. Put a tick in the upper box in for each combination of metal/solution where you expect a displacement reaction to take place. Alternatively, put a cross in the upper box for each combination where you don't expect a displacement reaction to take place.
3. Put about a 2 cm depth of each solution in four separate test tubes.
4. Add a different sample of metal to each tube.
5. Wait for about 5 minutes and then look for any changes to the metal or to the solution. Changes might include a coating on the metal or bubbles forming around the metal. You may need to tip the metal out of the solution and examine it closely to see any change.
6. Record your observations in the table below using the lower box for each combination.

Make observations and measurements

Metal	Solution of metal ions				
	Copper(II) sulfate solution $CuSO_4(aq)$	Magnesium sulphate solution $MgSO_4(aq)$	Iron(II) sulfate solution $FeSO_4(aq)$	Lead(II) nitrate solution $Pb(NO_3)_2(aq)$	Zinc sulfate solution $ZnSO_4(aq)$
Copper Cu					
Magnesium Mg					
Iron Fe					
Lead Pb					
Zinc Zn					

Analyse and interpret data

6. Why have some of the boxes in the table been shaded?
7. Which combinations did not match the prediction you made? Can you suggest a reason for this?
8. Write equations for all the displacement reactions which took place.

Activity C2f.3 Limiting corrosion

Learning objectives

- To consider the relative merits of different ways of preventing iron from rusting
- To undertake a devise and plan investigation

Learning outcomes

- Be able to describe that air (oxygen) and water must be present for iron to rust
- Be able to describe how rusting can be prevented by using grease, oil, plastic or by galvanising
- Understand the sacrificial protection of iron in terms of the reactivity series

Common misconceptions

Steel does not corrode (presumably confusion caused by knowledge of stainless steel).

Resources

Student Book page 155–157

Worksheet C2f.3a Preventing corrosion

Files on CD–ROM: C2f.3a_worksheet, C2f.3a_tech_notes

Resources for a class practical (see below)

Approach

Ask students to say under what conditions iron will rust. Ask why iron might rust more rapidly in sea water. Use the information in the Student Book page 155 to consider the different ways of protecting the corrosion or rusting of iron.

Ask students to work in small groups and to plan an investigation to compare the effectiveness of different ways of limiting the rusting of iron.

Worksheet C2f.3a provides a structure for students to plan and then when they put their plan into practice. The task is very straightforward. You will need to judge whether students need additional prompting about the things they should consider as they plan.

Students will need to consider the best ways of attaching the magnesium and zinc to the iron nails. It is likely that they will need to record observations after about 48 hours.

Technician's notes

You will need the following resources for a class practical.

Be sure to check the latest safety notes on these resources before proceeding.

shiny iron/mild steel nails
sodium chloride + spatula
grease
oil
magnesium ribbon
zinc foil
test tubes and test tube rack

Answers

Page 155

1. Air (oxygen) and water must be present.
2. The grease can be easily removed or wiped away.
3. a) Galvanizing involves coating iron or steel with zinc.

   b) As zinc is more reactive than iron, moist air will react with zinc in preference to the iron.

Page 157

1. a) $2Al(s) + Fe_2O_3(s) \rightarrow 2Fe(s) + Al_2O_3(s)$

   b) Aluminium is higher in the reactivity series than iron, therefore it is more reactive. Aluminium is able to displace the less reactive iron from its oxide and so form iron and aluminium oxide.

   c) Any metal which is higher than iron in the reactivity series can be selected. The higher the metal in the series, the more reactive metal and the more reactive the reaction will be. If the chosen metal is above aluminium, the reaction is more reactive.

   d) Aluminium is displacing iron in iron(III) oxide and becoming aluminium oxide by losing electrons. Iron(III) oxide is gaining electrons to become iron metal.

   Redox is when oxidation and reduction occur. Aluminium is losing electrons (oxidation). Iron(III) oxide is gaining electrons (reduction).
2. a) oxygen + water

   b) chromium protects iron from oxygen and water/used as it is shiny – good decorative effect.
3. Aluminium forms aluminium oxide which acts as a protective layer – does not react with the air.

C2f.3a Preventing corrosion

Iron and steel corrode or rust quite quickly when exposed to moist air. The rusting is an oxidation process and the iron is converted into iron(III) oxide. The iron(III) oxide is porous and so rusting will continue even after the surface is covered in rust.

Apparatus

shiny iron or steel nails

common salt, grease, oil, magnesium ribbon, zinc foil

test tubes and test tube rack

SAFETY INFORMATION
Magnesium ribbon is highly flammable.

Devise and plan investigations

1. Plan an investigation to find out which is the best method for preventing the corrosion or rusting of iron nails when in contact with air and salt water.

The methods of prevention you are going to test are:

A. Smearing the iron with grease

B. Coating the iron with oil

C. Putting the iron in contact with magnesium

D. Putting the iron in contact with zinc

2. Decide on your method and how you will record your results.

3. Carry out your method, analyse your results and then evaluate your approach.

Method

Write your method below.

____

____

____

____

____

____

____

____

____

Page 1 of 2

Make observations and measurements

Draw a table below to record your results.

Analyse and interpret data

What conclusions can you draw from your observations?

Evaluate data and methods

How would you rate the success of your plan and method? What problems did you encounter? How could the design have been improved?

Page 2 of 2

Activity C2f.4 Consolidation and summary

Learning objectives

- To review the learning points of the topic
- To test understanding through answering questions

Learning outcomes

- Be familiar with the knowledge and understanding summarised in the Student Book Checklist
- Be able to apply this knowledge and understanding by answering the End of Topic Questions

Resources

Student Book pages 158–159

Approach

Focus on the End of Topic Questions to test understanding of the requirements of the Checklist. The Developing Investigative Skills feature can be used to revise the work on displacement reactions if this was not used in activity C2f.3. The Science in Context feature relates to the extraction of metals from their ores, which is covered in Topic C5a *Extraction and uses of metals*.

Answers

End of Topic Questions mark scheme

Question	Correct answer	Marks
1	Sodium, calcium, magnesium, zinc, copper (minus 1 for each misplaced metal)	2 marks
2 a)	Mg(s) + PbO(s) → MgO(s) + Pb(s) (1 mark for correct formulae; 1 mark for balancing)	2 marks
2 b)	Magnesium	1 mark
2 c)	A redox reaction is one involving reduction and oxidation.	1 mark
3 a)	Y, Q, Z, X (minus 1 mark for each misplaced metal)	2 marks
3 b)	i) Y ii) X oxide iii) X oxide iv) Y	1 mark 1 mark 1 mark 1 mark
4 a)	Displacement reaction	1 mark
4 b)	Zinc is more reactive than copper.	1 mark
5 a)	Galvanising	1 mark
5 b)	The zinc is more reactive than the iron. The zinc reacts with oxygen in the air before the iron does.	1 mark 1 mark
5 c)	Painting, applying grease, plastic coating (1 mark for each)	3 marks
	Total:	20 marks

C2g Tests for ions and gases

Introduction

Although this may seem like a stand-alone topic, the terms will confuse students if they have not already studied Topic 1f *Ionic compounds*. Students will have already encountered the tests for oxygen, carbon dioxide and hydrogen if you teach this topic after Topic 2d *Oxygen and oxides* and Topic 2e *Hydrogen and water.*

The terms 'cation' and 'anion' link to Topic 1i *Electrolysis* (if this has not been studied). The topic focuses on tests for cations in solids and then in solutions. It includes the test for the ammonium ion and anion tests for carbonates, halides and the sulfate radical. It concludes with the tests for gases.

There are plenty of opportunities for students to practise and develop their investigative skills through qualitative analysis and identification activities.

Links to other topics

Sections	Essential background knowledge	Useful links
1 Principles of chemistry	1e Chemical formulae and chemical equations 1f Ionic compounds	1i Electrolysis
2 Chemistry of the elements	2d Oxygen and oxides 2e Hydrogen and water	2b Group 1 elements 2c Group 7 elements

Topic overview

C2g.1	**Orientation** This short introductory activity can be used to revise ideas on ionic bonding and the charges on ions. You may link this to the *Periodic Table* if this topic has been taught. In addition, you may revise some of the tests for gases already encountered. You may introduce the terms 'cation' and 'anion' if students are unfamiliar with this terminology.
C2g.2	**Identifying cations** This activity has three themes: the identification of metal ions in solids using a simple flame test technique; the identification of metal ions in solution using sodium hydroxide solution; and the identification of the ammonium ion. Worksheets C2g.2a *Using a flame test to identify metal ions* and C2g.2b *Identifying metal ions in solution* support student practical work.
C2g.3	**Identifying anions** This activity also has three themes: the identification of solid carbonates by reaction with dilute hydrochloric acid; the identification of the halide ions (chloride, bromide and iodide) using acidified silver nitrate solution; and the use of acidified barium chloride solution to identify the sulphate ion in solution. Worksheet C2g.3 *Identifying non-metal ions in solution* supports the identification of the halide and sulfate ions in solution.

C2g.4	**Identifying gases** This activity brings together information about individual confirmatory tests for gases, including tests for hydrogen, oxygen, carbon dioxide, chlorine and ammonia. Most of these tests should be familiar to students. Worksheet C2g.4 *Identifying gases* provides a key which supports a systematic approach to identifying the five gases.
C2g.5	**Consolidation and summary** This activity provides an opportunity for a quick recap of the ideas encountered in the topic and time for the students to answer the End of Topic Questions in the Student Book.

Activity C2g.1 Orientation

Learning objectives

- To recap on existing knowledge of ions and identification tests for gases

Learning outcomes

- Know that metals form positive ions which are called cations
- Know that non-metals form negative ions which are called anions
- Know the identification tests for hydrogen, oxygen and carbon dioxide

Resources

Student Book page 160

Approach

This is a short activity. Ask students to summarise in three or four points what they know about ionic bonding. If they have studied the Periodic Table, ask them how the table can help them to work out the charges on ions.

Explain that the topic has three main sections. The first looks at some tests that can be used to identify positive ions (also called cations), mainly metals ions but also the ammonium ion; the second section deals with tests for non-metal ions (also called anions). These tests together will identify the composition of some simple compounds.

If the *Electrolysis* topic has not yet been covered, explain that the terms 'cation' and 'anion' relate to how ions behave in electrolysis. This topic is covered in more detail later in the course.

The third section looks at identification tests for gases. Ask students to write down any gas identification tests they already know about. Explain that in this topic the following gases will be considered: hydrogen, oxygen, carbon dioxide, chlorine and ammonia. Students will probably not have encountered the test for ammonia.

Activity C2g.2 Identifying cations

Learning objectives

- To detect the presence of certain metal ions (cations) using a flame test
- To introduce the test for the ammonium ion
- To explore the use of sodium hydroxide solution to detect metal ions in solution

Learning outcomes

- Be able to describe a flame test using a nichrome wire, concentrated hydrochloric acid and a sample of a solid
- Know the flame colours produced by the following metal ions: Li^+; Na^+; K^+ and Ca^{2+}
- Be able to describe how to test for the ammonium ion, NH_4^+, using sodium hydroxide and know what indicates a positive result in each case
- Be able to describe tests for Cu^{2+}, Fe^{2+} and Fe^{3+} ions using sodium hydroxide solution and know what indicates a positive result in each case

Resources

Student Book pages 160–162

Worksheet C2g.2a Using a flame test to identify metal ions

Worksheet C2g.2b Identifying metal ions in solution

Files on CD–ROM: C2g.2a_worksheet, C2g.2b_worksheet, C2g.2a_tech_notes, C2g.2b_tech_notes, C2g.2c_tech_notes

Resources for a demonstration and a class practical (see below)

Approach

There are three parts to this activity:

1. Flame tests

This part is supported by Worksheet C2g.2a. Demonstrate the flame test technique, emphasising the importance of 'cleaning' the nichrome wire between tests and using the concentrated hydrochloric acid with care. Allow students to test the solids provided, including the unknown solids A and B.

2. Identifying metal ions in solution

This part is supported by Worksheet C2g.2b. The basis of the test is the precipitation of insoluble metal hydroxides. Calcium ions are included to provide a range of metal ions, but students are not required to remember the outcome of this test.

Again, students have to use their results to identify the metal ions present in two unknown solutions. Spend some time checking that they are able to write equations for these precipitation reactions.

3. Identifying the ammonium ion

Demonstrate the test using sodium hydroxide solution as described in the Student Book on page 162. Make the point that the ammonia gas is the only common alkaline gas and use of this fact will be made later in the topic when identifying gases.

Technician's notes

You will need the following resources for a class practical and a demonstration.

Be sure to check the latest safety notes on these resources before proceeding.

Part 1 – student practical
nichrome flame test wire in holder (cork or pyrex glass)
watch glasses, spatula, Bunsen burner
1M hydrochloric acid
copper(II) chloride, lithium chloride
calcium chloride, magnesium chloride, potassium chloride, sodium chloride
potassium sulfate (or other potassium salt), labelled as 'substance A'
calcium nitrate (or other calcium salt), labelled as 'substance B'
eye protection
Part 2 – student practical
beakers, droppers, test tubes, test tube rack
0.1M sodium hydroxide solution
0.1M solutions of calcium nitrate, copper(II) sulphate, iron(III) chloride
0.1M solutions of iron(II) ammonium sulfate (labelled iron(II) sulfate)
0.1M solutions of copper(II) chloride (labelled as 'solution M'), potassium chloride
0.1M iron(II) ammonium sulphate (labelled as 'solution N')
eye protection
Part 3 – demonstration
test tube, dropper, Bunsen burner
0.1M ammonium chloride solution
0.1M sodium hydroxide
universal indicator paper + chart

Note: iron(II) ammonium sulphate is used because the solution of iron(II) ions with ammonium ions is more stable than iron(II) sulphate.

Answers

Page 161

1. Calcium ions produce a brick red flame colour.
2. A nichrome wire is used in a flame test.
3. Add sodium hydroxide solution. Fe^{2+} produces a green precipitate; Fe^{3+} produces a reddish brown precipitate.

C2g.2a Using a flame test to identify metal ions

Some metal ions (cations) can be identified from the characteristic colours they produce when heated in a Bunsen flame. You will test a number of compounds containing known metal ions and then use your results to identify the metal ions present in solids A and B.

Apparatus

flame test wire and holder, watch glasses, spatula, Bunsen burner, 1M hydrochloric acid

compounds containing the following metal ions: calcium, copper, lithium, magnesium, potassium and sodium

compounds A and B

eye protection

SAFETY INFORMATION
1M hydrochloric acid is corrosive.
Copper(II) chloride and lithium chloride are harmful.
Wear eye protection.

Method

1. Put on your eye protection.

2. Heat a flame test wire in the hottest part of a Bunsen burner flame until the only colour produced in the flame is a faint orange.

3. Put a spatula measure of the solid you are going to test on a watch glass.

4. Dip the wire into some clean 1M hydrochloric acid on a watch glass and then into the solid sample. Put the flame test wire into the hottest part of the Bunsen burner flame. Record the colour of the flame produced in the table below.

5. Repeat the procedure for each of the solids you are testing. Remember to clean the wire after each test by heating it in the flame and to use fresh concentrated hydrochloric acid each time.

Make observations and measurements

Metal ion (cation)	Flame colour
calcium (Ca^{2+})	
copper(Cu^{2+})	
lithium (Li^{+})	
magnesium (Mg^{2+})	
potassium (K^{+})	
sodium (Na^{+})	
solid A	
solid B	

Analyse and interpret data

1. Identify the metal ions present in solid A and solid B.

C2g.2b Identifying metal ions in solution

Many metal ions (cations) can be detected in solution from the colour of the precipitate they form when sodium hydroxide solution is added. This test is only successful when the hydroxide of the metal ion is insoluble in water – that is, if it forms a precipitate in the solution.

Apparatus

beaker, dropper, sodium hydroxide solution

test tubes, test tube rack

solutions containing the following metal ions: calcium, copper, iron(II), iron(III) and potassium

solutions M and N

eye protection

SAFETY INFORMATION

Sodium hydroxide solution is an irritant.

Wear eye protection.

Method

1. Put on your eye protection.
2. Put about a 2 cm depth of the solution to be tested in a test tube.
3. Using a dropper, carefully add several drops of sodium hydroxide solution. Record your observations in the table below.

Make observations and measurements

Metal ion (cation) in solution	Observations on adding sodium hydroxide solution
calcium (Ca^{2+})	
copper (Cu^{2+})	
iron(II) (Fe^{2+})	
iron(III) (Fe^{3+})	
potassium (K^{+})	
solution M	
solution N	

Analyse and interpret data

1. Which metal ion did not produce a precipitate? Why was a precipitate not formed in this case?
2. Identify the metal ions present in solutions M and N.
3. Write chemical equations for the following reactions:
   a) Cu^{2+} ions (in copper(II) sulfate solution) and sodium hydroxide solution
   b) Fe^{3+} ions (in iron(III) chloride solution) and sodium hydroxide solution.

Activity C2g.3 Identifying anions

Learning objectives

- To identify the presence of the carbonate ion
- To identify certain non-metal ions in solution

Learning outcomes

- Be able to describe a test for CO_3^{2-} ions using dilute hydrochloric acid and know what indicates a positive result
- Be able to describe a test for Cl^-, Br^- and I^- ions using dilute nitric acid and silver nitrate solution and know the colours of the different precipitates
- Be able to describe a test for SO_4^{2-} ions using dilute hydrochloric acid and barium chloride solution and know what indicates a positive result

Resources

Student Book pages 162–164

Worksheet C2g.3a Identifying non-metal ions in solution

Files on CD–ROM: C2g.3a_worksheet, C2g.3a_tech_notes, C2g.3b_tech_notes

Resources for a demonstration and a class practical (see below)

Approach

There are three parts to the activity.

1. Identifying the carbonate ion

Students should be familiar with the test to identify carbon dioxide. Demonstrate the reaction of a carbonate, such as calcium carbonate, with dilute hydrochloric acid and bubble the gas through lime water. You may use apparatus similar to that on page 162 of the Student Book.

Give students time to write equations for this carbonate + acid reaction. Refer also to the equations given in the Student Book.

2. Identifying halide ions

Use Worksheet C2g.3a. In this part students test four unknown solutions A–D, record their observations and then identify the halide ion present using the information given on colours of halide precipitates.

Mention that the precipitates are light-sensitive and so may darken if left to stand.

3. Identifying sulfate ions

Use the second part of Worksheet C2g.3a. Students test four unknown solutions, record their results and then identify which solutions contain sulfate ions using the information provided on what constitutes a positive sulfate ion test.

Again, it is worth spending some time on writing chemical equations for these reactions.

Technician's notes

You will need the following resources for a demonstration and class practical.

Be sure to check the latest safety notes on these resources before proceeding.

Part 1 – demonstration
test tube, bung and delivery tube, test tube
lime water
Parts 2 & 3 – student practical
test tubes, test tube rack, droppers
0.05M silver nitrate solution, 0.1M nitric acid
0.1M potassium iodide (labelled as 'solution A'), 0.1M sodium chloride (labelled as 'solution B')
0.1M sodium bromide (labelled as 'solution C'), 0.1M sodium iodide (labelled as 'solution D')
0.1M sodium chloride (labelled as 'solution E'), 0.1M sodium sulfate (labelled as 'solution F')
0.1M potassium sulfate (labelled as 'solution G'), 0.1M sodium chloride (labelled as 'solution H')
0.1M barium chloride solution
0.1M hydrochloric acid
eye protection

Answers

Page 164

1. Add dilute sodium hydroxide and heat. An alkaline gas (turns red litmus paper blue) indicates the presence of an ammonium compound.

2. a) Carbon dioxide

   b) Bubble the gas through lime water. A white precipitate forms.

3. The Fe^{3+} ion is present in solution X.

4. The Cl^- ion is present in solution Y.

C2g.3a Identifying non-metal ions in solution

You need to be able to identify four non-metal ions or anions in solution:

a) The halide ions (Group 7): chloride, bromide and iodide

b) The sulfate ion

You will need to use two separate tests. In each case you will be given some solutions containing unknown non-metal ions and will have to use your observations and the information provided to identify the non-metal ions.

Apparatus

test tubes, test tube rack, droppers

silver nitrate solution and dilute nitric acid

barium chloride solution and dilute hydrochloric acid

solutions A, B, C, D, E, F, G and H

eye protection

SAFETY INFORMATION

Barium chloride solution is harmful.

Wear eye protection.

Method

1. Put on your eye protection.

Test 1

2. Put approximately a 2 cm depth of the test solution in test tube. Add approximately a 1 cm depth of dilute nitric acid followed by approximately a 1 cm depth of silver nitrate solution. Record your observations in Table 1.

Test 2

3. Put approximately a 2 cm depth of the test solution in test tube. Add approximately a 1 cm depth of dilute hydrochloric acid followed by approximately a 1 cm depth of barium chloride solution. Record your observations in Table 2.

Make measurements and observations

Table 1

Test Solution	Observations on adding dilute nitric acid and silver nitrate solution	Non-metal ion (anion) present
A		
B		
C		
D		

Page 1 of 2

Table 2

Test Solution	Observations on adding dilute hydrochloric acid and barium chloride solution	Is the sulfate ion (anion) present?
E		
F		
G		
H		

Analyse and interpret data

3. Use the information in the table below to identify the non-metal ions present in Test 1. Complete the last column of Table 1 above.

Non-metal ion	Observations on adding dilute nitric acid followed by silver nitrate solution
Chloride (Cl^-)	A white precipitate forms
Bromide (Br^-)	A cream coloured precipitate forms
Iodide (I^-)	A yellow precipitate forms

4. Write a chemical equation for the reaction of solution B with silver nitrate solution (You can assume that solution B is a solution of a sodium compound).

5. Use the information in the table below to identify which solutions contain the sulphate ion. Complete the last column of Table 2.

Non-metal ion	Observations on adding dilute hydrochloric acid followed by barium chloride solution
Sulfate (SO_4^{2-})	A white precipitate forms

6. Write a chemical equation for the reaction of solution F with barium chloride solution (You can assume that solution F is a solution of a sodium compound).

Page 2 of 2

Activity C2g.4 Identifying gases

Learning objectives

- To revise some simple identification tests for common gases

Learning outcomes

- Be able to describe tests for the gases hydrogen, oxygen, carbon dioxide, ammonia and chlorine and know what indicates a positive result in each case

Resources

Student Book page 164–166

Worksheet C2g.4a Identifying gases

Files on CD–ROM: C2g.4a_worksheet, C2g.4a_tech_notes

Resources for a class practical (see below)

Approach

This activity provides confirmation or revision for the individual tests for the gases in this topic. Students should be familiar with these from previous work. Worksheet C2g.4a provides a key for identifying some samples of unknown gases.

Technician's notes

You will need the following resources for a class practical.

Be sure to check the latest safety notes on these resources before proceeding.

universal indicator paper and chart, splints, Bunsen burner
lime water
samples of gases (labelled A–F) in corked test tubes:
A – hydrogen
B – carbon dioxide
C – ammonia
D – chlorine
E – oxygen
F – carbon dioxide
eye protection

Answers

Page 165

1. Ammonia
2. Oxygen
3. Chlorine

Page 166

1. a) Dip wire in concentrated hydrochloric acid, heat strongly in flame until absence of any yellow colour (sodium)

   b)

Name of cation	Colour of precipitate
Zinc/lead	white
Magnesium/calcium	white
Copper(II)	blue
Iron(II)	green/turns brown slowly
Iron(III)	rust brown/orange

   c) $Ag^+(aq) + X^-(aq) \rightarrow AgX(s)$ where X^- is Cl^-, Br^-, I^-.

   d) HCl is added to remove any carbonate ions that may be present.

2. Plan needs to check for testing of both anion and cation for each sample and should include practical instructions.

   Blue compound

   Test for copper(II) – sodium hydroxide: result blue precipitate

   $Cu^{2+}(aq) + 2OH^-(aq) \rightarrow Cu(OH)^2(s)$

   Test for sulphate – hydrochloric acid/barium chloride: result white precipitate

   $Ba^{2+}(aq) + SO_4^{2-}(aq) \rightarrow BaSO_4(s)$

   White compound

   Flame test for Na^+ – yellow

   Test for carbonate – add dilute acid – effervescence/carbon dioxide evolved – turns lime water milky

   $CO_3^{2-}(s) + 2H^+(aq) \rightarrow H_2O(l) + CO_2(g)$

C2g.4a Identifying gases

You need to know the identification tests for the following gases:

hydrogen oxygen carbon dioxide ammonia chlorine

You will be provided with samples of unknown gases to identify.

Apparatus

universal indicator paper, splints, lime water solution

access to gas samples A, B, C, D, E and F

eye goggles

SAFETY INFORMATION

Chlorine (if used) is toxic and dangerous to the environment.

Wear eye goggles.

Method

1. Put on your eye goggles.
2. Use the key below to identify the gases. Record your observations in the table.

Identification key

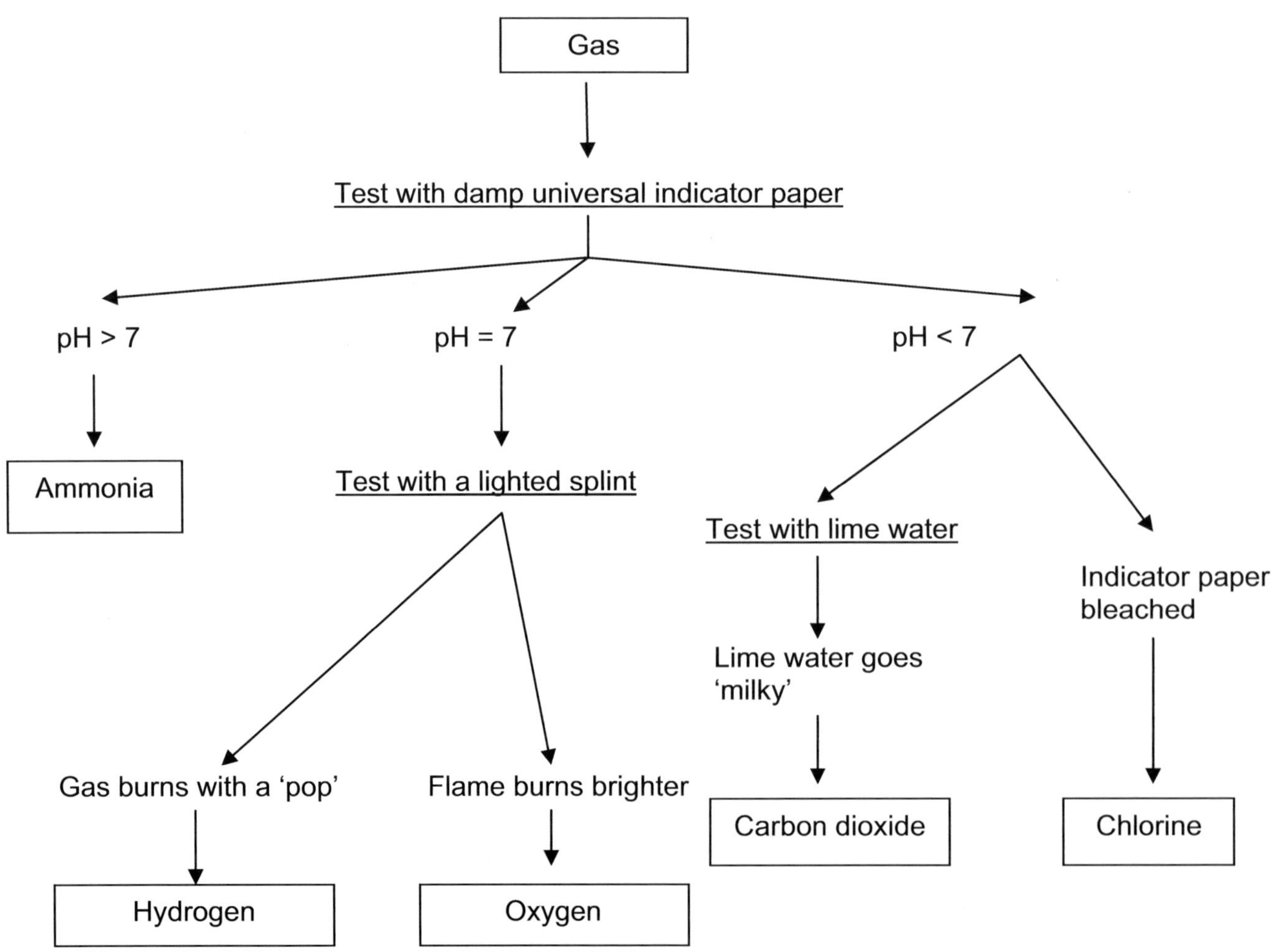

Make observations and measurements

Gas	Result of tests used	Gas identification
A		
B		
C		
D		
E		
F		

Analyse and interpret data

1. Use your observations to identify the gas present in each sample. Complete the last column of the table above.
2. Write an equation for the reaction that occurs when a lighted splint is applied to hydrogen.
3. Write an equation for the reaction between carbon dioxide and lime water.

Activity C2g.5 Consolidation and summary

Learning objectives

- To review the learning points of the topic
- To test understanding through answering questions

Learning outcomes

- Be familiar with the knowledge and understanding summarised in the End of Topic Checklist
- Be able to apply this knowledge and understanding by answering End of Topic Questions

Resources

Student Book page 167–168

Approach

Focus on the End of Topic Questions to test students' understanding of the Checklist. Use the Science in Context feature on flame emission spectroscopy to emphasise an important application of the identification of metal ions using a flame test technique. The topic lends itself to a '20 questions' approach to revision or consolidation. Use questions such as:

- What is the test for hydrogen?
- What observation indicates a positive result in the identification of sulfate ions?

Answers

End of Topic Questions mark scheme

Question	Correct answer	Marks
1 a)	Use a nichrome wire. Dip the wire into concentrated hydrochloric acid, then the solid, then the flame.	1 mark 1 mark
1 b)	Sodium compounds produce a yellow/orange flame. Potassium compounds produce a lilac/purple flame.	1 mark 1 mark
2 a)	Cu^{2+} ions.	1 mark
2 b)	Ammonium ions, NH_4^+	1 mark
2 c)	Fe^{3+} ions	1 mark
3	Completed table as shown below	3 marks

Gas	Test	Observations
Chlorine	Damp universal indicator paper	*Indicator paper bleached/goes white*
Carbon dioxide	Bubble through lime water	White precipitate of suspension forms
Hydrogen	*Apply a lighted splint*	Burns with a 'pop'

4	Use a flame test (nichrome wire, concentrated hydrochloric acid). Lithium ions produce a bright red flame. Add dilute hydrochloric acid and bubble any gas through lime water. A carbonate will produce carbon dioxide which will turn the lime water cloudy.	1 mark 1 mark 1 mark 1 mark
5 a)	Add dilute hydrochloric acid. Then barium chloride solution. A white precipitate indicates the presence of a sulfate ion.	1 mark 1 mark 1 mark
5 b)	Add dilute nitric acid. Then silver nitrate solution. A yellow precipitate indicates the presence of the iodide ion.	1 mark 1 mark 1 mark
	Total:	20 marks

Section 2: Exam-style questions mark scheme

Question	Correct answer	Marks
2 a)	Si	1 mark
2 b)	S	1 mark
2 c)	Group 1	1 mark
2 d)	Group 2	1 mark
2 e)	Group 7	1 mark
3 a)	2, 8, 7	1 mark
3 b)	7	1 mark
3 c)	Brown fumes to colourless gas	1 mark
3 d)	i) Red hydrogen iodide forms an acid in water ii) Green hydrogen iodide will not form an acid unless dissolved in water – it would be neutral	1 mark 1 mark 1 mark 1 mark
4 a)	Magnesium, zinc, iron	1 mark
4 b)	Magnesium + sulfuric acid → magnesium sulfate + hydrogen	1 mark
4 c)	$Mg(s) + H_2SO_4(aq) \rightarrow MgSO_4(aq) + H_2(g)$	1 mark
4 d)	Copper (silver, gold)	1 mark
4 e)	The reaction with water/steam or the reaction with oxygen	1 mark
5 a)	Use nichrome wire, dip into concentrated hydrochloric acid Then the solid, then in the flame	1 mark 1 mark
5 b)	Sodium carbonate Sodium hydrogen carbonate	1 mark 1 mark
5 c)	Any two of: lithium carbonate, calcium carbonate, lithium hydroxide, calcium hydroxide	1 mark for each
6 a)	i) Displacement/redox reaction ii) $Mg(s) + PbO(s) \rightarrow MgO(s) + Pb(s)$ iii) Magnesium. It displaces lead from lead(II) oxide/it reduces the lead(II) oxide to lead	1 mark 2 marks 1 mark
6 b)	i) Magnesium ii) Lead(II) oxide iii) Lead(II) oxide iv) Magnesium	1 mark 1 mark 1 mark 1 mark
6 c)	i) $Mg(s) + Pb(NO_3)_2(aq) \leftrightarrow Mg(NO_3)_2(aq) + Pb(s)$ ii) Any metal above Pb in the reactivity series, e.g. aluminium iii) Any metal below lead in the reactivity series, e.g. copper	2 marks 1 mark 1 mark

Section 3. Organic Chemistry

Contents

a) Alkanes

b) Alkenes

c) Ethanol

d) Exam-style questions

Overview of the section

This section covers the organic chemistry content of the course. There will be more opportunities to apply this learning in Section 5. This section concentrates on the three homologous series of alkanes, alkenes and alcohols – in the case of alcohols; the content is limited to a study of ethanol.

It is unlikely that students have studied organic chemistry before starting this course. You will need to spend time developing an understanding of naming in organic chemistry and improving their familiarity with terms such as 'saturated', 'unsaturated', 'addition' and 'isomerism'.

All the topics in the section depend on a clear understanding of atomic structure, covalent bonding and the characteristic properties of simple covalent compounds. Discussion of the combustion of hydrocarbons links to the *Oxygen and oxides* and *Hydrogen and water* topics. The topic on alkanes links directly to the crude oil topic in Section 5, and the topic on alkenes links directly to the synthetic polymers topic in Section 5.

In the suggested teaching sequence, all the topics are included in the second year of the course, and you can link them relatively easily to the relevant topics in Section 5. As with other sections, it is important to use orientation to establish the extent of your students' prior knowledge.

Starting points

The Student Book provides a 'section opener', a double page spread which sets the scene for the students (see pages 174–175). It is structured in this way:

- an overview providing details of the areas of study
- six questions, the first three relating to aspects of chemistry that should be reasonably familiar to most students, and the final three relating directly to some of the topics to be studied
- a list of section contents.

The questions provide a structure for introducing the section.

1. Where is carbon in the Periodic Table of elements? What can you work out about carbon from its position?

2. What is the atomic number of carbon? How are its electrons arranged?

3. How does carbon form covalent bonds? Can you show the bonding in methane (CH_4), the simplest of organic molecules?

4. You will be learning about series of organic compounds which are hydrocarbons. What do you think a hydrocarbon is?

5. You will be learning about methane. Do you know where methane can be found and what it is used for?

6. You will also be learning about ethanol, which belongs to a particular series of organic compounds. Do you know where you could find ethanol in everyday products?

The section opener has two main purposes:

- to acknowledge students' prior learning and to value it
- to provide a benchmark against which future learning can be compared.

The six questions or 'starting points' in the Student Book can be used in a number of different ways to introduce the section:

- You could ask students to consider the questions as an introductory homework task.
- You could put students into groups to share their own ideas and understanding and then report back to the whole class.
- Students could be given access to the Internet, preferably with a tight timescale, to find out the information required.

You could then use a spider chart or other form of wall chart to summarise everybody's ideas.

Recording these initial ideas in this way allows you to retain them for reference as the individual topics are developed. In this way, your students' progress in learning can be readily acknowledged.

C3a Alkanes

Introduction

This is the first topic in the Organic Chemistry section of the specification. Students should already be familiar with electronic structure (Topic 1c *Atomic structure*) and covalent bonding (Topic 1g *Covalent substances*). This topic provides a basis for the subsequent study of Topic 3b *Alkenes* and Topic 3c *Ethanol*, as well as for Topic 5b *Crude oil* and Topic 5c *Synthetic polymers*.

The topic establishes the nature of a homologous series and introduces terms such as 'hydrocarbon' and 'saturated', followed by alkanes as the basis for naming all organic compounds. You will revise dot and cross diagrams and introduce structural isomers. Combustion and substitution reactions are highlighted as the main reactions of alkanes. This topic does not provide opportunities for investigative work.

Links to other topics

Sections	Essential background knowledge	Useful links
1 Principles of chemistry	1c Atomic structure 1g Covalent substances	
2 Chemistry of the elements	2e Hydrogen and water	
3 Organic chemistry		3b Alkenes 3c Ethanol
5 Chemistry in society		5b Crude oil 5c Synthetic polymers

Topic overview

C3a.1	**Orientation** This activity starts with a revision of covalent bonding (you may use Worksheet C3ab.1 *Covalent bonding in alkanes*) and introduces the concept of a homologous series within a new branch of chemistry – organic chemistry. The names of the first five alkanes are introduced, together with information about their states of matter at room temperature and their uses as fuels. The term 'hydrocarbon' is introduced.
C3a.2	**The structure of alkanes** This is a short activity which focuses on the existence of structural isomers for alkanes with four or more carbon atoms.
C3a.3	**The chemical properties of alkanes** This activity starts with the combustion of alkanes, emphasising their general use as fuels. The concept of complete and incomplete combustion is introduced, and the potential dangers associated with incomplete combustion. Substitution is then introduced as another reaction of alkanes, with the reaction between methane and bromine as the example.
C3a.4	**Consolidation and summary** This activity provides an opportunity for a quick recap of the ideas encountered in the topic and time for the students to answer the End of Topic Questions in the Student Book.

Activity C3a.1 Orientation

Learning objectives

- To revise the nature of covalent bonding
- To introduce the concept of a homologous series with alkanes as an example

Learning outcomes

- Know that a homologous series is a group of organic compounds with the same general formula, similar chemical properties and physical properties which change gradually from one member to the next
- Know that alkanes have a general formula of C_nH_{2n+2}
- Know that a hydrocarbon is a compound containing carbon and hydrogen only

Resources

Student Book page 176

Worksheet C3a.1a Covalent bonding in alkanes

Files on CD–ROM: C3a.1a_worksheet, C3a.1a_tech_notes

Resources for demonstration (see below)

Approach

Explain that this topic is from a branch of chemistry called organic chemistry, and that alkanes are a group or homologous series of organic chemicals.

Explain the characteristics of a homologous series: general formula, similar chemical properties and a trend in physical properties.

Show samples of alkanes: methane (natural gas/Bunsen burner), propane (camping gas canister), butane (lighter fuel) and pentane. Mention that they are all fuels and their physical properties vary from methane as a gas, through to propane and butane as gases that are often liquefied under pressure, to pentane a liquid. Introduce and explain the term 'hydrocarbon'.

Ask students what they can recall about the nature of covalent bonding. Use molecular models to illustrate the structure of alkanes and Worksheet C3ab.1 to recap on dot and cross bonding diagrams.

Show how the formulae of alkanes match the general formula C_nH_{2n+2}.

Technician's notes

You will need the following resources for a demonstration.

Be sure to check the latest safety notes on these resources before proceeding.

a bottle or canister of camping gas (propane)
lighter fuel (butane)
bottle of pentane

C3a.1a Covalent bonding in alkanes

Methane, ethane and propane are compounds in the alkane homologous series. They are covalently bonded.

Draw dot and cross diagrams showing the bonding in each of these compounds. In each case the displayed formula has been inserted in the table.

(Atomic numbers: H = 1; C = 6)

Alkane	Dot and cross diagram	Displayed formula
Methane		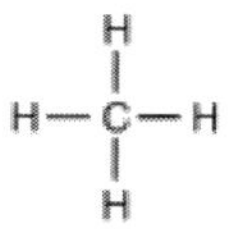
Ethane		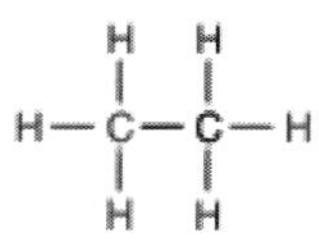
Propane		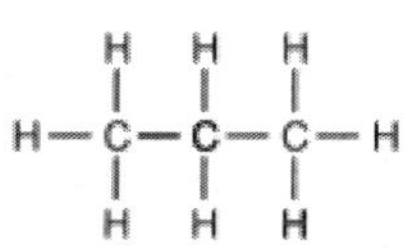

Activity C3a.2 The structure of alkanes

Learning objectives

- To look in more detail at the structure of alkanes, including non-straight chain isomers
- To introduce the concept of a saturated/unsaturated hydrocarbon

Learning outcomes

- Be able to name the first five straight-chain members of the alkane homologous series (methane, ethane, propane, butane and pentane)
- Be able to draw displayed formulae for the first five members of the alkane homologous series
- Know that isomerism is the existence of compounds with the same molecular formulae but different structural formulae
- Know that a saturated hydrocarbon contains only carbon to carbon single bonds (whereas an unsaturated hydrocarbon contains C=C double bonds)

Resources

Student Book pages 176–180

Approach

Explain that so far alkanes have been shown as straight-chain molecules. Introduce the concept of isomers and branched-chain molecules with the same molecular formulae.

Show molecular models of the butane and 2-methylpropane. Explain how the names of these isomers relate to the number of carbon atoms in the longest chain.

Page 180 of the Student Book shows the three structural isomers of pentane. Emphasise the fact that although the molecules have the same formula, they have different structures and therefore different physical properties, as exemplified by the different boiling points of the isomers of pentane.

One of the questions in the Student Book on page 180 refers to hexane and some of its isomers. All the isomers contain C-C single bonds – explain that this type of molecule is said to be 'saturated' (students may be familiar with this term in the context of saturated and unsaturated fats in food). Mention that unsaturated hydrocarbons contain C=C double bonds (see the next topic).

Answers

Page 179

1. a) Contains no C=C double bonds

   b) A compound containing hydrogen and carbon only

2. a) $C_{15}H_{32}$

   b) Carbon dioxide and water

Page 180

1. Isomers are molecules with the same molecular formula but different structural formulae.

2.

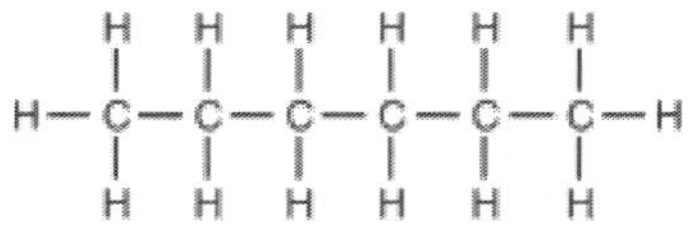

The molecular formula is C_6H_{14}.

3. Any two of the following:

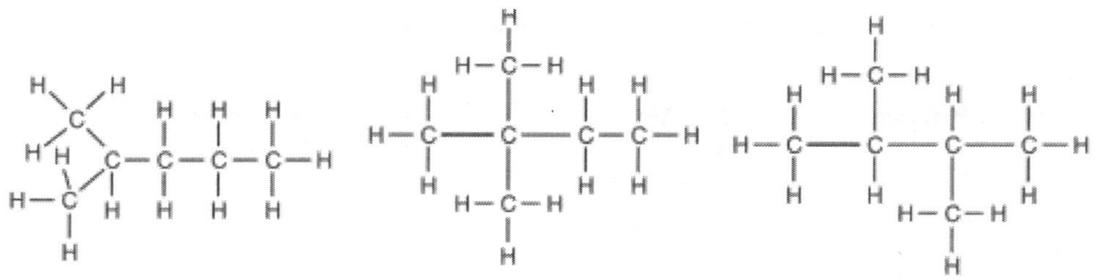

4. Hexane will be a liquid. Its physical properties will most closely resemble those of pentane.

Activity C3a.3 The chemical properties of alkanes

Learning objectives

- To study the combustion reactions of alkanes and the nature of the products formed
- To study the substitution reactions of alkanes and the nature of the products formed

Learning outcomes

- Know that when alkanes burn in a plentiful supply of air or oxygen they form carbon dioxide and water
- Know that in a limited supply of air or oxygen (incomplete combustion) alkanes can burn to form carbon and carbon monoxide
- Know that methane will react with bromine in the presence of UV light to form bromomethane (CH_3Br)

Resources

Student Book pages 181–182

Approach

Students will be familiar with the combustion of alkane fuels. Explain that hydrocarbons burn in a plentiful supply of air or oxygen to form carbon dioxide and water (as exemplified by the blue flame on a Bunsen burner).

Ask what happens if the air supply is restricted (the yellow flame on the Bunsen burner). Emphasise that incomplete combustion can produce carbon and carbon monoxide, and the danger caused by carbon monoxide.

Give students practice at writing equations for the complete and incomplete combustion of simple hydrocarbons. Refer to the equations on page 181 of the Student Book.

Then introduce students to the substitution reaction of alkanes. Emphasise what conditions are needed and why the reaction is called a substitution reaction. Refer to page 181 of the Student Book.

Activity C3a.4 Consolidation and summary

Learning objectives

- To review the learning points of the topic
- To test understanding through answering questions

Learning outcomes

- Be familiar with the knowledge and understanding summarised in the End of Topic Checklist
- Be able to apply this knowledge and understanding by answering the End of Topic Questions

Resources

Student Book pages 182–184

Approach

Focus on the End of Topic Questions to test students' understanding of the requirements of the End of Topic Checklist.

Refer to the *Science in Context* feature on methane to highlight the fact that methane is a 'greenhouse' gas and some scientists think that it could have an influence on global warming.

Answers

End of Topic Questions mark scheme

Question	Correct answer	Marks
1 a)	A homologous series is group of organic compounds with the same general formula, similar chemical properties and physical properties which change gradually from one member to the next.	1 mark
1 b)	Structural isomerism is the existence of compounds with the same molecular formula but different structural formulae.	1 mark
2	$C_{10}H_{22}$	1 mark
3	No. This compound does not have the general formula of C_nH_{2n+2}.	1 mark
4 a)	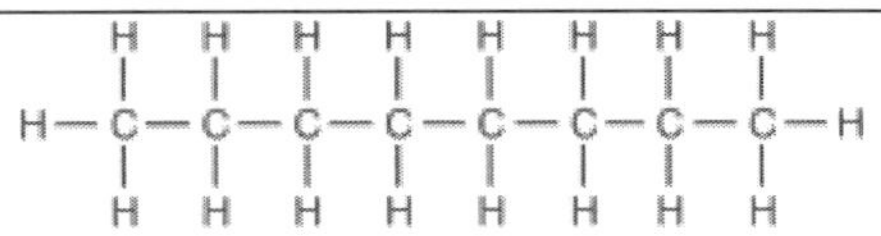	1 mark
4 b)	Diagram must have 8 carbon atoms and 18 hydrogen atoms with at least one branch. Some examples are shown here. 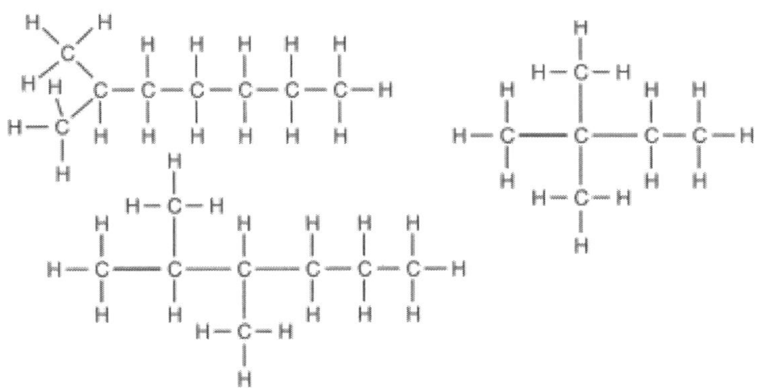	1 mark
4 c)	Liquid	1 mark
4 d)	Octane is a saturated hydrocarbon – it contains only C-C single bonds (no C=C bonds).	1 mark
5 a)	C_2H_6.	1 mark
5 b)	Carbon dioxide Water	1 mark 1 mark
5 c)	Blue flame	1 mark
5 d)	$2C_2H_6(g) + 7O_2(g) \rightarrow 4CO_2(g) + 6H_2O(l)$ (1 mark for formulae; 1 mark for balancing)	2 marks
5 e)	Carbon Carbon monoxide	1 mark 1 mark
5 f)	Yellow flame	1 mark
6 a)	Substitution	1 mark
6 b)	In the presence of ultra violet (UV) light.	1 mark
6 c)	Bromomethane	1 mark
	Total:	20 marks

C3b Alkenes

Introduction

This is a short topic that extends understanding of covalent bonding (Topic 1g *Covalent substances*) and the introductory work on alkanes (Topic 3a *Alkanes*). It links to the production of alkenes by the catalytic cracking of alkanes in Topic 5b *Crude oil* and to the formation of addition polymers in Topic 5c *Synthetic polymers*.

The topic focuses on two aspects of alkenes: their structures and their chemical reactions, including combustion and the addition reaction with bromine and bromine water. There are few opportunities for investigative work, but students will be able to use bromine water to identify an alkene in the next topic (Topic 3c *Ethanol*).

Links to other topics

Sections	Essential background knowledge	Useful links
1 Principles of chemistry	1c Atomic structure 1g Covalent substances	
3 Organic chemistry	3a Alkanes	3c Ethanol
5 Chemistry in society		5b Crude oil 5c Synthetic polymers

Topic overview

C3b.1	**Orientation** This activity picks up on many of the ideas covered in the *Alkanes* topic: the characteristics of a homologous series and the general formula for an alkene; the distinction between a saturated and unsaturated hydrocarbon; and the covalent bonding associated with a C=C double bond.
C3b.2	**The chemical properties of alkenes** This activity focuses on the combustion of alkenes as typical hydrocarbons and the addition reaction with bromine. The decolourisation of bromine water is introduced as a way of identifying an alkene as an unsaturated hydrocarbon.
C3b.3	**Consolidation and summary** This activity recaps on the ideas encountered in the topic, as well as considering a *Science in Context* feature on 'Saturated and unsaturated fats' in the Student Book. Time can be allocated to answering the End of Topic Questions in the Student Book.

Activity C3b.1 Orientation

Learning objectives

- To recap knowledge of covalent bonding and the alkane homologous series
- To introduce the names and displayed formulae for simple alkenes

Learning outcomes

- Know that alkenes have a general formula of C_nH_{2n}
- Know that unsaturated hydrocarbons contain C=C double bonds
- Be able to name the first three straight chain members of the alkene homologous series (ethene, propene, butene)
- Be able to draw displayed formulae for the first three members of the alkene homologous series, including the isomers of butene

Resources

Student Book page 185

Approach

Ask students to name the first four members of the alkane homologous series and to draw the displayed formulae for these.

Ask students if they can remember the characteristics of a homologous series: general formula, gradual trend in physical properties and common chemical properties. Explain that alkenes have a general formula C_nH_{2n} and ask them to draw the displayed formula for propene (C_3H_6). This should illustrate that the covalent bonding must involve a C=C double bond, that is, the alkenes are unsaturated hydrocarbons.

Use Table 3.4 on page 185 of the Student Book to look at the trend in boiling points, and the subsequent states of matter at room temperature and pressure, of the alkene series.

Introduce the structural isomers that exist for butene, and the system for naming them by numbering the carbon atoms in the straight chain (for example, but-1-ene and but-2-ene).

Activity C3b.2 The chemical properties of alkenes

Learning objectives

- To look at the combustion of alkenes as an homologous series of hydrocarbons
- To introduce the reaction of an alkene with bromine as an addition reaction

Learning outcomes

- Know that alkenes are hydrocarbons and so will burn in a plentiful supply of air to form carbon dioxide and water
- Understand what an addition reaction with an alkene involves
- Be able to describe the reaction of an alkene with bromine water and know that this reaction can be used to identify alkenes

Resources

Student Book pages 185–188

Files on CD–ROM: C3b.3a_tech_notes

Resources for demonstration (see below)

Approach

Revise the combustion of hydrocarbons and the products formed in a plentiful supply of air (carbon dioxide and water) and in a limited supply of air (carbon and carbon monoxide). What observation would suggest that the hydrocarbon was burning in a limited supply of air? (A yellow flame)

Ask students to write chemical equations for the combustion of ethene in both a plentiful supply of air and in a limited supply of air (forming carbon and water).

Demonstrate the reaction of an alkene (hexene or cyclohexene) with bromine water. The bromine water is decolourised as an addition reaction takes place. Use Fig. 3.10 on page 186 of the Student book to illustrate the displayed formula of the product of the reaction and the reason for describing the reaction as 'addition'.

Explain that this reaction is used to identify alkenes (or more accurately any unsaturated hydrocarbon, including alkynes, which contain a C≡C triple bond). Tell students that they will have an opportunity to use this identification test later in the course.

Technician's notes

You will need the following resources for a demonstration.

Be sure to check the latest safety notes on these resources before proceeding.

an alkene (such as hexene or cyclohexene)
bromine water
test tube, droppers, test tube holder

Answers

Page 187

1. It contains at least one C=C double bond.

2. The manufacture of polymers (polyethene).

3.

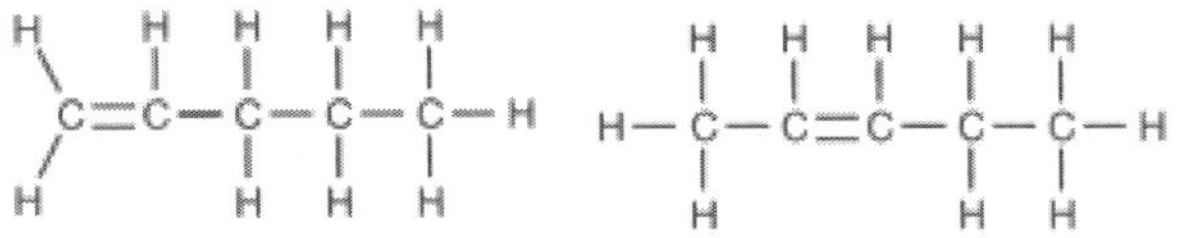

Page 188

1. a) Saturated compounds contain only covalent single bonds. Alkanes contain only carbon–carbon and carbon–hydrogen single bonds and are therefore saturated hydrocarbons. (Alkenes contain C=C double bond and are therefore unsaturated.)

   b) Alkane: C_6H_{14}, alkene C_6H_{12}. The position of the double bond can be between any pair of carbon atoms, but there should be one less hydrogen attached to the double-bonded carbons.

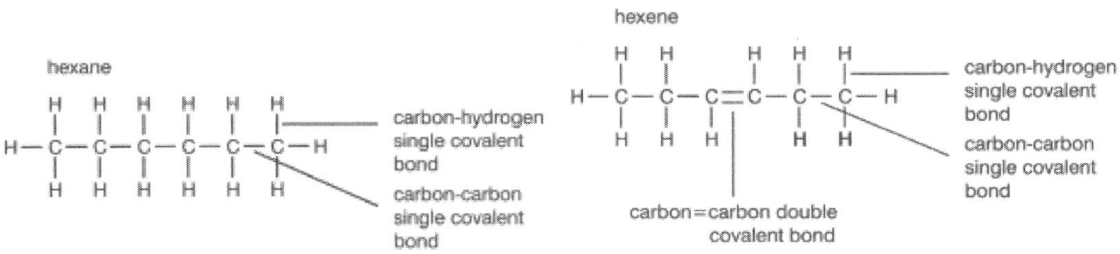

   c) Isomers of hexane – need to have 5 carbon atoms in a straight chain with a carbon as a branch or 4 carbons in a line and two branches (a straight chain should not be written with a bend).

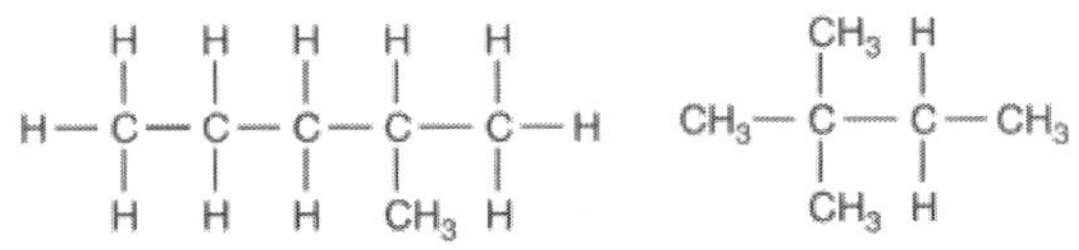

   d) Hexane undergoes substitution reaction with bromine (any H can be substituted with Br/can have more than one substitution). HBr is also formed in the reaction. The diagram shows 1-bromohexane.

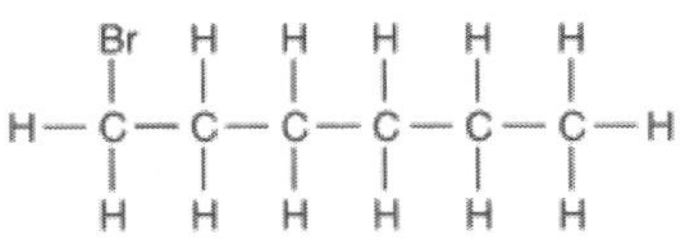

   Hexene undergoes an addition reaction with bromine (note bromine loses its colour – this is a test for unsaturation). The bromine will add across the double bond. Example shows addition of hex-1-ene to form 1,2 dibromohexane.

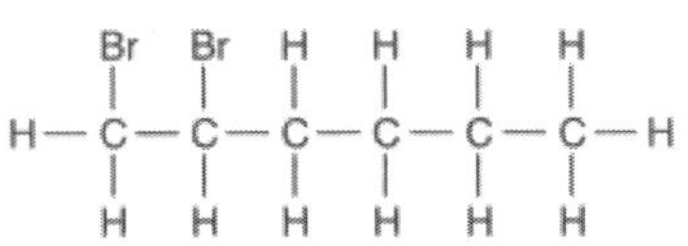

2. a) Fuels are substances that provide heat energy. Alkanes burn readily in air combining with oxygen to produce carbon dioxide and water vapour and large quantities of heat.

   b) Incomplete combustion leads to the formation of carbon monoxide instead of carbon dioxide. It is a very poisonous gas and particularly dangerous as it has no odour and causes drowsiness. Carbon monoxide is poisonous because it reacts with the haemoglobin in the blood, forming 'carboxyhaemoglobin'. The haemoglobin is no longer available to carry oxygen to the body and death results from oxygen starvation.

   c) Short-chain hydrocarbons are more likely to form carbon dioxide and water as their main products as there is less carbon per molecule in these hydrocarbons to react with the available oxygen. Longer-chain hydrocarbons often burn with a smoky flame and leave black carbon deposits as there is insufficient oxygen to form carbon dioxide, with the many carbons in the longer chains. Carbon monoxide is also formed.

Activity C3b.3 Consolidation and summary

Learning objectives

- To review the learning points of the topic
- To test understanding through answering questions

Learning outcomes

- Be familiar with the knowledge and understanding summarised in the End of Topic Checklist
- Be able to apply this knowledge and understanding by answering the End of Topic Questions

Resources

Student Book pages 187–190

Approach

Ask students to answer the End of Topic Questions and study the End of Topic Checklist.

Use the *Science in Context* feature on 'Saturated and unsaturated fats' on pages 187–188 to emphasise the importance of recognising this distinction in a healthy diet.

Answers

End of Topic Questions mark scheme

Question	Correct answer	Marks
1 a)	Yes. It has the general formula of an alkene – a series of unsaturated hydrocarbons (contain C=C double bonds).	1 mark
1 b)	Heptene	1 mark
1 c)	Isomers of heptene: diagrams should have seven carbon atoms with a double bond between two carbon atoms as shown below. 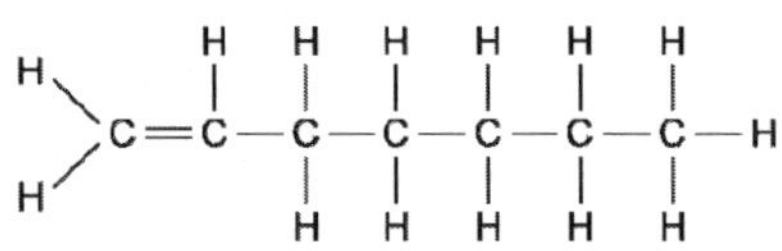 and 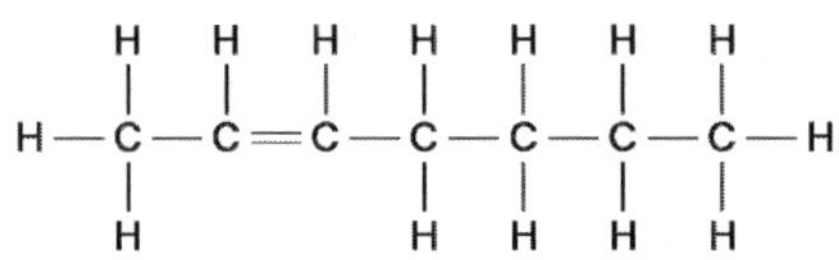	2 marks
2 a)	C_8H_{16}.	1 mark
2 b)	Liquid – pentene with 5 carbon atoms is a liquid so octane is likely to be a liquid.	1 mark
3 a)	Carbon dioxide Water	1 mark 1 mark
3 b)	i) $C_2H_4(g) + 3O_2(g) \rightarrow 2CO_2(g) + 2H_2O(l)$ (1 mark for formulae; 1 mark for balancing) ii) Blue flame	2 marks 1 mark
3 c)	i) $C_2H_4(g) + O_2(g) \rightarrow 2C(g) + 2H_2O(l)$ (1 mark for formulae; 1 mark for balancing) ii) Yellow flame	2 marks 1 mark
4 a)	Diagrams of hexane (left) and hex-l-ene (right) e.g. 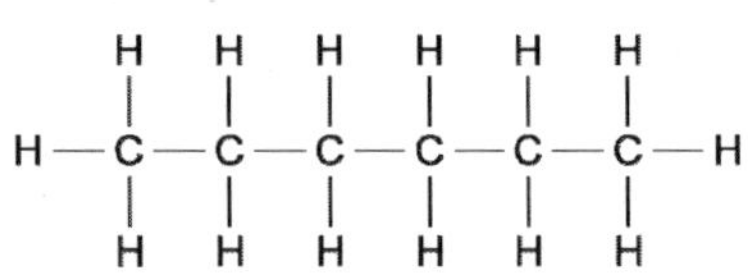 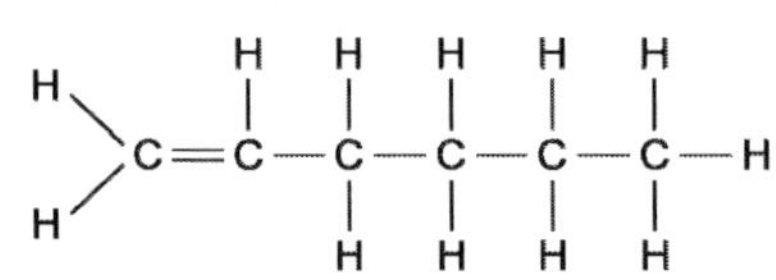	2 marks
4 b)	Hexene	1 mark
4 c)	Bromine water	1 mark
5 a)	Orange To colourless/decolourised	1 mark 1 mark
5 b)	$C_3H_6(g) + Br_2(aq) \rightarrow C_3H_6Br_2(l)$ (1 mark for formulae; 1 mark for balancing)	2 marks
5 c)	An addition reaction	1 mark
6	Unsaturated fat	1 mark
	Total:	24 marks

C3c Ethanol

Introduction

This topic introduces alcohols, the third and final homologous series in the *Organic Chemistry* section, although the focus is almost exclusively on ethanol. There are few direct links to topics studied later in the course other than to the production of ethene by the catalytic cracking of crude oil (Topic 5b *Crude oil*).

As with Topic 3a *Alkanes* and Topic 3b *Alkenes*, covalent bonding in alcohols can be linked back to Topic 1g *Covalent substances*, although you may not need to emphasise this link if this topic follows on closely after *Alkanes* and *Alkenes*.

The topic builds on the concept of a homologous series and compares the manufacture of ethanol by fermentation, including the use of fractional distillation to separate the ethanol from water, with that from ethene and steam. The merits and problems of using ethanol as a fuel are considered. After that, students have an opportunity to make ethene by the dehydration of ethanol, using bromine water to identify the unsaturated hydrocarbon produced.

Links to other topics

Sections	Essential background knowledge	Useful links
1 Principles of chemistry	1c Atomic structure 1g Covalent substances	
3 Organic chemistry	3a Alkanes 3b Alkenes	
5 Chemistry in society		5b Crude oil

Topic overview

C3c.1	**Orientation** This short activity revises the concept of a homologous series and introduces the general formula for the alcohol series. The focus quickly centres on ethanol, often referred to simply as 'alcohol', and some of its important uses.
C3c.2	**The manufacture of ethanol** This activity is in three parts. The first part involves the manufacture of ethanol by fermentation and the use of fractional distillation to separate ethanol from water. The second part considers the manufacture of ethanol from ethene, obtained from crude oil. Finally, the two methods are compared, along with the factors that might determine the preferred method of manufacture. The Developing Investigative Skills box on page 195 of the Student Book could be used here to consolidate understanding of fermentation, or it could be used as revision in Activity C3c.4 at the end of the topic.
C3c.3	**Reactions of ethanol** Initially the activity focuses on the combustion of ethanol and its use as a fuel for motor cars. The Science in Context feature on page 197 of the Student Book is used to stimulate discussion of the issues arising from using the fermentation of sugar to generate a fuel. Students then have the opportunity to undertake the dehydration of ethanol in order to collect and test the ethene produced.

C3c.4	**Consolidation and summary** This activity recaps on the key ideas encountered in the topic. Time may be allocated to answering the End of Topic Questions in the Student Book. Developing Investigative Skills could be used as a revision if not already used in Activity C3c.2.

Activity C3c.1 Orientation

Learning objectives

- To introduce alcohols as another homologous series of organic compounds
- To consider the covalent bonding in the alcohols

Learning outcomes

- Know that alcohols have a general formula of $C_nH_{2n+1}OH$
- Know that ethanol is the commonest alcohol

Common misconceptions

Some students think that there is only one alcohol. It is a single substance and not the name of a homologous series.

Resources

Student Book pages 191–193

Approach

Introduce alcohols as the final homologous series covered in the course. Provide the general formula and ask students to work out the molecular formula, and then draw the displayed formula, of the first four alcohols of the series.

Demonstrate that propanol and butanol have structural isomers by asking students to draw different structures for the two alcohols.

Use page 192 of the Student Book to consider the common uses of ethanol.

Answers

Page 193

1. C_4H_9OH

2. It is a relatively 'clean' fuel and releases only carbon dioxide and water into the atmosphere. (It does not release sulfur dioxide and nitrogen oxides, as petrol does when it burns.)

3. A solvent is a liquid that dissolves other substances (solutes) to form solutions.

Activity C3c.2 The manufacture of ethanol

Learning objectives

- To introduce two methods of manufacturing ethanol
- To consider the factors that influence the choice of method used

Learning outcomes

- Be able to describe the fermentation process used in the manufacture of ethanol from sugar, including the importance of temperature and the use of yeast
- Be able to describe the manufacture of ethanol using ethene and steam passed over a catalyst of phosphoric acid, with a temperature of 300 °C and pressure of 70 atmospheres
- Be able to evaluate the factors relevant to the choice of method used in the manufacture of ethanol, including the relative availability of sugar cane and crude oil

Common misconceptions

The ethanol produced by fermentation is different (and more healthy) than that produced from ethene.

Resources

Student Book pages 193–196

Files on CD–ROM: C3c.2a_tech_notes

Resources for demonstration (see below)

Approach

The activity conveniently splits into three parts:

1. Fermentation demonstration

Set up a demonstration of the fermentation method for making ethanol from sugar, using apparatus similar to that shown on page 195 of the Student Book. Alongside this, demonstrate how fractional distillation can be used to separate the ethanol from water after fermentation is complete. Use apparatus similar to that shown on page 194 of the Student Book and emphasise that the collected product can be determined from the temperature of the thermometer at the top of the column (ethanol has a boiling point of 78 °C). The ethanol can be identified by burning a small quantity on a pyrex watch glass or in a crucible. You could use Developing Investigative Skills on page 195 to consolidate understanding of the fermentation process. Alternatively, you might use it as a revision task at the end of the topic.

2. Manufacture of ethanol

Use the Student Book page 196 to illustrate the conditions required to make ethanol from ethene (which is obtained by cracking fractions obtained from crude oil). Remind students that this is an example of an addition reaction which they encountered in Topic C3c *Alkenes*. Emphasise that the extreme conditions required (300 °C and 70 atmospheres pressure) mean that the production costs are relatively high.

3. Which method of manufacture?

Use the table on page 196 of the Student Book to stimulate a discussion about the factors that determine which method of manufacture is chosen. (The use of sugar as a renewable source is also considered in the next activity in the context of: 'Is this the best way of using a foodstuff such as sugar?')

Technician's notes

You will need the following resources for a demonstration.

Be sure to check the latest safety notes on these resources before proceeding.

Part 1
simple fermentation apparatus: conical flask, water bath, bung and delivery tube, test tube (page 195 in Student Book)
sugar (glucose), yeast, limewater
fractional distillation apparatus: Bunsen burner and heatproof mat, clamps and stands, circular flask, fractionating column containing glass beads, condenser and tubes (page 194 in Student Book)
ethanol
anti-bumping granules, eye protection, Bunsen burner

Answers

Page 194

1. Fermentation is the process in which ethanol is made from sugar, yeast and water.
2. The optimum temperature is in the range 25 to 30 °C.
3. The yeast contains enzymes which increase the rate of the reaction.
4. Carbon dioxide

Developing investigative skills, page 195

1. Use a thermostatically controlled heater in the water bath or leave the apparatus near to a heater to maintain the temperature within the range needed or heat gently with a Bunsen burner or add hot water to maintain temperature
2. To show that gas was being produced (by the bubbles) and to identify the gas as carbon dioxide (turns the lime water cloudy).
3. Frothing in the flask/bubbles in the limewater/limewater turning milky.
4. No more bubbles through the lime water.
5. Ethanol. The boiling point of ethanol is 76 °C which is below that of water (100 °C).

Activity C3c.3 Reactions of ethanol

Learning objectives

- To consider the advantages of using ethanol as a fuel
- To introduce the preparation of ethene by the dehydration of ethanol

Learning outcomes

- Understand the advantages of using ethanol as a fuel
- Be able to describe the dehydration of ethanol to ethene using aluminium oxide

Resources

Student Book pages 196–198

Worksheet C3c.3a The dehydration of ethanol

Files on CD–ROM: C3c.3a_worksheet, C3c.3a_tech_notes

Resources for demonstration and class practical (see below)

Approach

Students may be familiar with the use ethanol in spirit burners. If not, you can demonstrate this use. Ask students to write the chemical equation for the complete combustion of ethanol.

If necessary, emphasise the names of the products formed when ethanol burns in a limited supply of air. Use the Science in Context feature on page 197 of the Student Book to support a discussion of the issues associated with using ethanol obtained by fermentation as a fuel.

The second part of the activity is a student practical involving the dehydration of ethanol. Details of the practical work are included on Worksheet C3d.3a. It is important to emphasise the dangers of a 'suck-back' if heating is stopped while the delivery tube is still in the water in the trough.

Students use bromine water to detect ethene as an unsaturated hydrocarbon.

Technician's notes

You will need the following resources for a demonstration and a class practical.

Be sure to check the latest safety notes on these resources before proceeding.

Demonstration
spirit burner

Student practical
Bunsen burner and heatproof mat, clamp and stand, 2 boiling tubes, bung and piping, water trough
mineral wool, tongs, wooden splints, Bunsen burners
ethanol
bromine water
aluminium oxide
eye protection

Answers

Page 198

1. a) $CH_2=CH_2(g) + H_2O(g) \rightleftharpoons C_2H_5OH(g)$ ($\Delta H = -45$ kJ/mol)

Any reaction in which heat is given out by a system to the surroundings is called an exothermic reaction.

b) The reaction for the formation of ethanol is reversible. The forward reaction is exothermic and therefore low temperatures are favoured in order to produce maximum product (that is, to shift the position of equilibrium to the right). Low temperatures however slow down the rate of the reaction. Manufacturers are trying to produce as much ethanol as possible each day and therefore a compromise temperature is used to balance the amount of ethanol produced with the time it takes. High temperatures, although would increase the rate would produce a poor yield of ethanol.

c) In the reaction for the formation of ethanol from ethene and steam there are two molecules on the left hand side of the equation but only one on the right, therefore the formation of ethanol is favoured by high pressure. High pressures also increase the rate of reaction, but high pressures are expensive as the plant needs strong pipes and containment vessels and high energy requirements. High pressure also causes the ethene to polymerise. Pressures of 60–70 atmospheres are therefore used.

d) Although adding phosphoric(V) acid as a catalyst does not produce any greater percentage of ethanol in the equilibrium mixture, it does ensure that the reaction is fast enough for the dynamic equilibrium to be set up within the time the gases are within the reactor. Without the catalyst the reaction would be too slow to make the manufacture of ethanol cost-effective.

C3c.3a The dehydration of ethanol

Ethene can be prepared by the dehydration of ethanol. The reaction takes place if ethanol vapour is passed over very hot aluminium oxide. The ethene can be collected by displacement of water.

Apparatus

Bunsen burner and heatproof mat

clamp and stand

2 boiling tubes

bung and delivery tube

bowl

mineral wool

tongs

ethanol, aluminium oxide and bromine water

eye protection

SAFETY INFORMATION

Bromine water is harmful.

Ethanol is highly flammable.

Handle the mineral wool with tongs – do not use your fingers.

Wear eye protection.

Method

1. Put on your eye protection.
2. Set up the apparatus as shown in the diagram. The mineral wool should absorb all the ethanol used – add a few drops.

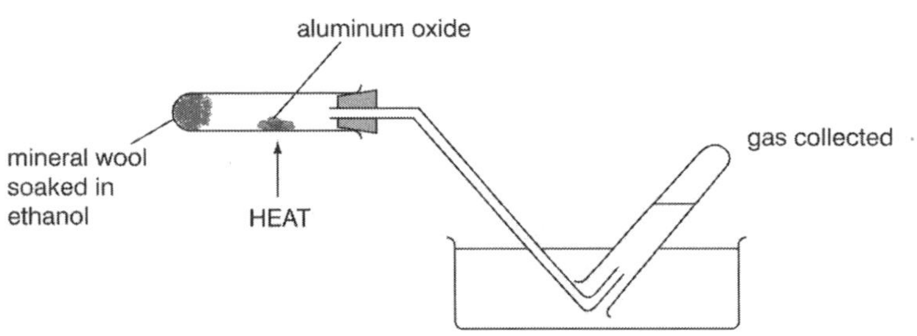

3. Heat the aluminium oxide strongly and occasionally flick the flame onto the mineral wool.
4. Collect at least 3 test tubes full of gas.
5. **Before stopping heating, remove the delivery tube from the trough of water.**
6. Test separate test tubes of the gas collected with:
   a) a lighted splint
   b) a small quantity of bromine water.

Page 1 of 2

Make observations and measurements

Record the results of your tests in the table.

Test	Observations
1. A lighted splint added to the gas	
2. Bromine water added to the gas	

Analyse and interpret data

1. What is the name given to this type of reaction?
2. Write an equation for the preparation of ethene from ethanol.
3. Is the gas collected saturated or unsaturated? Explain your answer.

Evaluate data and methods

1. Why might the first test tube of gas collected not contain ethene?
2. Why should the delivery tube be removed from the trough of water before heating is stopped?

Activity C3c.4 Consolidation and summary

Learning objectives

- To review the learning points of the topic
- To test understanding through answering questions

Learning outcomes

- Be familiar with the knowledge and understanding summarised in the End of Topic Checklist
- Be able to apply this knowledge and understanding by answering End of Topic Questions

Resources

Student Book pages 199–200

Approach

The Developing Investigative Skills feature on fermentation could be used to revise the ideas encountered in Activity C3c.2, if it was not used in that activity.

Ask students to study the End of Topic Checklist and then answer the End of Topic Questions.

Answers

End of Topic Questions mark scheme

Question	Correct answer	Marks
1	$C_nH_{2n+1}OH$	1 mark
2 a)	$C_5H_{11}OH$	1 mark
2 b)	Pentanol: 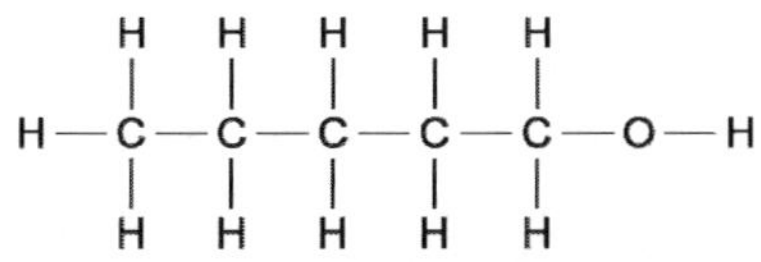	1 mark
2 c)	i) An isomer is a molecule with the same molecular formula as another molecule but different structural formula. ii) Pentan-3-ol 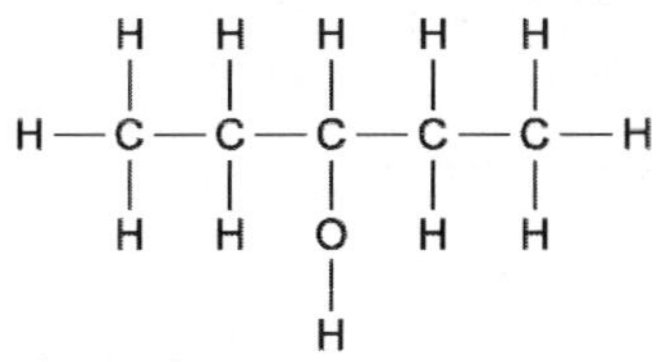	1 mark 1 mark
3	Sugar cane Sugar beet	1 mark 1 mark
4 a)	Enzymes present in the yeast speed up the fermentation process Enzymes are denatured at temperatures above about 35 °C	1 mark 1 mark
4 b)	Fermentation is an anaerobic process and takes place in the absence of oxygen. If oxygen were present the ethanol would be oxidised (to ethanoic acid or vinegar).	1 mark 1 mark
5 a)	Ethene is obtained from crude oil.	1 mark
5 b)	The phosphoric acid is a catalyst.	1 mark
5 c)	Temperature 300 °C Pressure 60-70 atmospheres	1 mark 1 mark
5 d)	$C_2H_4(g) + H_2O(g) \rightarrow C_2H_5OH(g)$	1 mark
6 a)	Fermentation uses renewable resources/ethanol has suitable flavour for alcoholic drinks.	1 mark
6 b)	The method involving ethene is a much faster process than fermentation.	1 mark
7 a)	Aluminium oxide catalyst Heat/high temperature	1 mark 1 mark
7 b)	$C_2H_5OH(l) \rightarrow C_2H_4(g) + H_2O(l)$	1 mark
7 c)	Dehydration	1 mark
	Total:	22 marks

Section 3: Exam-style questions mark scheme

Question	Correct answer	Marks
2 a)	i) True ii) False iii) True iv) False v) False Subtract 1 mark for each mistake, but no negative marks.	2 marks
2 b)	i) A compound/molecule containing carbon and hydrogen only ii) A compound/molecule containing only C to C single bonds	1 mark 1 mark
2 c)	i) C_3H_8 ii) Diagram as shown in Table 3.1 on page 177 of Student Book	1 mark 2 marks
3 a)	C_6H_{12}	1 mark
3 b)	Liquid	1 mark
3 c)	55–75 °C	1 mark
3 d)	i) Add bromine water ii) The bromine water will be decolourised iii) No reaction	1 mark 1 mark 1 mark
3 e)	Diagrams as shown in Fig. 3.11 on page 187 of the Student Book.	2 marks
3 f)	i) Ethanol ii) Phosphoric acid iii) $C_2H_4(g) + H_2O(g) \rightarrow C_2H_5OH(g)$	1 mark 1 mark 2 marks
4 a)	i) Fermentation ii) Dehydration	1 mark 1 mark
4 b)	i) Temperature 25–30 °C ii) Yeast as a catalyst	1 mark 1 mark
4 c)	$C_2H_5OH(g) \rightarrow C_2H_4(g) + H_2O(g)$	1 mark
4 d)	i) Phosphoric acid catalyst ii) 300 °C iii) 60–70 atmospheres pressure	1 mark 1 mark 1 mark
4 e)	Diagram as shown in Table 3.6 on page 191 of the Student Book	1 mark

Contents

a) Acids, alkalis and salts

b) Energetics

c) Rates of reaction

d) Equilibria

e) Exam-style questions

Overview of the section

This section covers most of the physical chemistry aspects of the course, a discussion continued in Section 5 *Chemistry in society*. The topic *Acids, alkalis and salts* starts with the simple identification of acids and alkalis using indicators, then progresses to definitions based on proton donation and proton acceptance. Practical methods for the preparation of soluble and insoluble salts are then considered.

The topic on energetics considers exothermic and endothermic classifications, focusing on calorimetric techniques for measuring energy changes. Extension work covers making and breaking of bonds and the associated average bond energies. The *Rates of reaction* topic uses the simple collision theory to explain the observed impact of concentration, temperature, surface area and catalysts on the rate of a reaction. A short topic on equilibria completes the section, introducing the concept of a dynamic equilibrium along with a discussion of how reaction conditions affect the yield of a chemical reaction.

All the topics give students a chance to practise and perfect the use of equations to describe chemical reactions. In addition, the topic on acids, alkalis and salts builds on previous work on ions and ionic compounds. The topic on energetics topic provides an opportunity to revise ideas on covalent bonding. The concepts developed in the topics on rates of reaction and equilibria relate directly to the study of the Haber and Contact processes in the topic in Section 5 on the industrial manufacture of chemicals. In the same way, the topic on energetics feeds into the Section 5 topic on crude oil.

In the suggested teaching sequence, the topics are spread out over both years of the course. The first three topics in particular provide a range of experimental and investigatory activities which can be used to develop and extend your students' practical skills. As with other sections, it is important to use the 'orientation' stage to establish the extent of their prior knowledge.

Starting points

The Student Book provides a 'section opener', a double page spread which sets the scene for the students (see pages 204–205). It is structured in this way:

- an overview providing details of the areas of study
- six questions, the first three relating to aspects of chemistry that should be reasonably familiar to most students, and the final three relating directly to some of the topics to be studied
- a list of section contents.

The questions provide a structure for introducing the section.

1. How can you easily distinguish between an acid and an alkali?
2. How many non-renewable fuels can you name? What products do they form when they burn?
3. Can you give an example of a very rapid, almost instantaneous, chemical reaction? Can you give an example of a very slow one?
4. Have you heard of the process of neutralisation? If so, in what context?
5. Do you know, or could you find out, what a catalyst is? Do you know of any examples where catalysts are used in everyday life?
6.Can you give any examples of processes that are reversible?

The section opener has two main purposes:

- to acknowledge the student's prior learning and to value it
- to provide a benchmark against which future learning can be compared.

The six questions or 'starting points' in the Student Book can be used in a number of different ways to introduce the section:

- You could ask students to consider the questions as an introductory homework task.
- You could put students into groups to share their own ideas and understanding and then report back to the whole class.
- Students could be given access to the Internet, preferably with a tight timescale, to find out the information required.

You could then use a spider chart or other form of wall chart to summarise everybody's ideas.

Recording these initial ideas in this way allows you to retain them for reference as the individual topics are developed. In this way, your students' progress in learning can be readily acknowledged.

C4a Acids, alkalis and salts

Introduction

This topic builds on knowledge of ions (introduced in Topic 1f *Ionic compounds)* and mole calculations using volumes and molar concentrations (in Topic 1e *Chemical formulae and chemical equations*). You will find opportunities for reinforcement of acids and bases in studying the alkali metals in Topic 2b *Group 1 elements*; the conditions under which hydrogen chloride behaves as an acid in Topic 2c *Group 7 elements*; and the acidic/basic nature of oxides in Topic 2d *Oxygen and oxides*.

There are other links through neutralisation and enthalpy change in Topic 4b *Energetics* and the industrial manufacture of sulfuric acid and sodium hydroxide in Topic 5d *The manufacture of chemicals*.

The topic starts with the use of indicators to identify acids, alkalis and neutral substances. This is likely to be a revision for most students. You will then consider the characteristics of acids, bases and alkalis before concentrating on soluble and insoluble salts, their nature and methods of preparation. Students have good opportunities to develop their practical investigative skills, including using the titration technique accurately. They also perform a mole calculation to determine the concentration of dilute hydrochloric acid using the results of a titration with sodium hydroxide solution.

Links to other topics

Sections	Essential background knowledge	Useful links
1 Principles of chemistry	1e Chemical formulae and chemical equations 1f Ionic compounds	
2 Chemistry of the elements		2b Group 1 elements 2c Group 7 elements 2d Oxygen and oxides
4 Physical chemistry		4b Energetics
5 Chemistry in society		5d The industrial manufacture of chemicals

Topic overview

C4a.1	**Orientation** This activity seeks to revise and consolidate ideas on acids and alkalis, using indicators to classify them. A number of indicators are used to emphasise the benefits of a universal indicator. The pH scale facilitates the classification of solutions as strongly or weakly acidic or alkaline.
C4a.2	**Acids, bases, alkalis and salts** This activity considers the nature of acids, bases, alkalis and salts. Acids are defined as sources of H^+ ions and alkalis as sources of OH^- ions. Salts are introduced as the substances produced when the acidic hydrogen(s) of an acid is/are replaced by a metal.
C4a.3	**Making soluble salts** This activity starts with considering four methods for producing soluble salts, and then focuses on the practical aspects of two methods. First, the reaction between an acid and an alkali and the titration technique – this is extended to consider mole calculations involving volumes and concentrations. The second focuses on the preparation of hydrated magnesium sulfate by the reaction between magnesium carbonate and dilute sulfuric acid.

	Both preparation techniques provide students with opportunities for developing their investigative skills. The Developing Investigative Skills feature on the titration technique could be used as part of this activity, or it could be used as revision in Activity C4a.5 later in the topic.
C4a.4	**Making insoluble salts** This activity uses the rules provided in the Student Book for determining the solubility or insolubility of salts. Students perform a series of test tube reactions to familiarise themselves with some precipitation reactions. They are also given practice at writing chemical equations for these reactions.
C4a.5	**Consolidation and summary** This activity recaps on the key ideas encountered in the topic. The Developing Investigative Skills feature could be used as a means of revising the titration technique if it has not been used in Activity C4a.3. Time can be allocated to answering the End of Topic Questions in the Student Book.

Activity C4a.1 Orientation

Learning objectives

- To revise the use of indicators to classify substances as acidic, neutral or alkaline
- To compare different indicators
- To reinforce the use of universal indicator to identify substances which are strongly or weakly acidic or alkaline

Learning outcomes

- Know how indicators such as litmus, phenolphthalein and methyl orange can be used to distinguish between acidic and alkaline solutions
- Know the colours of each of the above indicators in acidic and alkaline solutions
- Know how universal indicator can be used to test a solution
- Understand how the pH scale from 0 to 14 can be used to classify solutions as strongly acidic, weakly acidic, neutral, weakly alkaline or strongly alkaline

Common misconceptions

Take care how you use the terms 'strong' and 'weak' in relation to acids and alkalis. The concept of strong and weak acids and alkalis (that is, the extent of dissociation into ions) is beyond the scope of this specification. To avoid confusion later, it is helpful to use terms such as 'strongly acidic' rather than 'strong acid', etc.

Resources

Student Book pages 206–207

Worksheet C4a.1a Using indicators

Files on CD–ROM: C4a.1a_worksheet, C4a.1a_tech_notes

Resources for class practical (see below)

Approach

Your students are very likely to have existing experience and knowledge of acid and alkalis, and will probably have used indicators such as litmus and universal indicators. This activity will revise and consolidate their ideas.

Give them an overview of the topic: it provides a more detailed study of the nature of acids and alkalis as well as focusing on a class of compounds called salts.

Ask students to list their current knowledge of acids, alkalis and indicators. Take feedback and summarise on a chart. Explain that they are going to undertake a straightforward practical activity to consolidate their understanding, possibly including some indicators they are not be familiar with – methyl orange and phenolphthalein.

Ensure that students understand the meaning of the terms 'strongly acidic', 'weakly acidic', 'neutral', 'weakly alkaline' or 'strongly alkaline'.

Use Worksheet C4a.1a *Using indicators* to support students in testing some 'everyday' chemicals. The advantages of using universal indicator and the pH scale will become obvious – it allows differentiation into strongly and weakly acidic or alkaline.

Technician's notes

You will need the following resources for a class practical.

Be sure to check the latest safety notes on these resources before proceeding.

test tubes, test tube rack
0.1M sodium hydroxide solution
0.1M hydrochloric acid
0.1M solution of bicarbonate of soda (sodium hydrogen carbonate)
soapy water, milk of magnesia, household 'ammonia' cleaner
grapefruit juice, vinegar, distilled water
methyl orange, phenolphthalein, universal indicator + chart
eye protection

C4a.1a Using indicators

Substances can be classified as acidic, alkaline or neutral. The colour an indicator shows in a solution of the substance can be used to classify the substance.

Apparatus

test tubes, test tube rack

dilute hydrochloric acid, dilute sodium hydroxide, distilled water, soapy water, grapefruit juice, bicarbonate of soda, milk of magnesia, household ammonia cleaner, vinegar

methyl orange, phenolphthalein, universal indicator paper or solution + chart

eye protection

SAFETY INFORMATION

Sodium hydroxide solution is an irritant.

Wear eye protection.

Method

Part 1

1. Put on your eye protection.
2. Test the dilute hydrochloric acid, distilled water and dilute sodium hydroxide with the indicators shown in the table. Record your observations.

Test solution	Colour of indicator in solution		
	Methyl orange	Phenolphthalein	Universal indicator (pH)
Dilute hydrochloric acid (A typical acid)			
Distilled water (Neutral)			
Dilute sodium hydroxide (A typical alkali)			

Page 1 of 2

Part 2

2. Test the solutions shown in the table with the indicators. Record your observations in each case.

Test solution	Colour of indicator when added to the test solution			Acid/alkali/ neutral/pH
	Methyl orange	Phenolphthalein	Universal indicator (pH)	
Soapy water				
Grapefruit juice				
'Bicarbonate'				
Milk of magnesia				
Household 'ammonia' cleaner				
Vinegar				

Analyse and interpret data

1. Complete the last column in your table of observations.
2. Which solution is:
   a) the most strongly acidic ______________________
   b) the most strongly alkaline ______________________
   c) neutral or the closest to being neutral? ______________________

Evaluate data and methods

1. What are the advantages of using universal indicator rather than methyl orange or phenolphthalein?

____

____

____

Page 2 of 2

Activity C4a.2 Acids, bases, alkalis and salts

Learning objectives

- To emphasise the nature of acids, bases, alkalis and salts, including hydrated salts
- To introduce the characteristics of acids and alkalis in terms of the ions they provide in solution

Learning outcomes

- Know that acids when in aqueous solution are a source of H^+ ions
- Know that alkalis when in aqueous solution are a source of OH^- ions
- Know that salts made from hydrochloric acid are called chlorides
- Know that salts made from nitric acid are called nitrates
- Know that salts made from sulfuric acid are called sulfates

Resources

Student Book pages 207–209

Approach

This is a relatively short activity which defines acids, bases, alkalis and salts. Use the section in the Student Book to introduce the definitions, paying particular attention to acids being a source of H^+ ions and alkalis a source of OH^- ions.

Your students may need a quick revision of ionic bonding at this point. Spend some time clarifying the nature of a salt and the names of salts linked to the acid they are formed from. Explain that some salts are hydrated and contain water of crystallization – show how this water of crystallization is represented in the formula. (Students may have encountered the terms before if they have used anhydrous copper(II) sulfate to identify the presence of water.)

Answers

Page 209

1. Both solutions are alkalis. Solution A is a weakly alkaline whereas solution B is a strongly alkaline.

2. The solution is acidic.

3. The basicity of an acid is tells you how many hydrogen ions can be replaced (for example in a reaction with a base).

4. Calcium is a metal. The oxides (and hydroxides) of metals are bases.

Activity C4a.3 Making soluble salts

Learning objectives

- To provide some simples rules for deciding whether a salt is soluble or insoluble
- To explore methods for preparing soluble salts, including the reaction between an acid and an alkali using the titration technique

Learning outcomes

- Be able to predict the products of reactions between dilute hydrochloric acid, nitric and sulfuric acids and metals, metal oxides and metal carbonates
- Be able to predict whether a simple salt is soluble or insoluble in water
- Be able to predict whether a carbonate is soluble or insoluble in water
- Know how soluble salts can be prepared by adding a metal, base or carbonate to an acid
- Know how a soluble salt can be prepared by titration using an acid and an alkali
- Be able to describe how to carry out an acid – alkali titration
- Understand how to calculate the concentration of an acid or alkali from the results of a titration experiment

Resources

Student Book pages 209–214, 216–217

Worksheet C4a.3a Making a soluble salt (1) – using a titration method

Worksheet C4a.3b Making a soluble salt (2)

Files on CD–ROM: C4a.3a_worksheet, C4a.3b_worksheet, C4a.3a_tech_notes, C4a.3b_tech_notes

Resources for class practical (see below)

Approach

Introduce the rules for whether a salt is soluble or insoluble. Explain that the focus in this activity will be the preparation of soluble salts.

Look at the four general methods as listed on page 211 of the Student Book. Explain to students that they will undertake some investigative work focused on Methods 1 and 3.

1. Acid and alkali

Students can use Worksheet C4a.3a to investigate Method 1 (acid + alkali). The focus is on the experimental technique: students don't actually go on and produce a sample of a salt, although the worksheet indicates how this would be done. Instead, focus on using the titration technique to work out the concentration of hydrochloric acid.

Emphasise the importance of working carefully – the titration technique can give an accuracy of one drop. Explain that the first set of results should be considered as a trial. It can be done relatively quickly to avoid drop by drop addition over a long period of time. Also show how to safety use the equipment i.e. filling and fitting fillers etc.

Your students are likely to need to perform at least two further titrations. Many students find calculations involving concentrations of solutions difficult to grasp, so you will need to spend some time on this. The worksheet encourages a step-by-step approach to aid understanding.

You could use the Developing Investigative Skills box on pages 216–217 of the Student Book to reinforce ideas on the titration technique, or alternatively this could be used at the end of the topic as part of Activity C4a.5.

2. Acid and carbonate

Students can use Worksheet C4a.3b to investigate Method 3 (acid + carbonate) by preparing a sample of hydrated magnesium sulfate. The method replicates that shown in the Student Book and may be applied to any soluble salt preparation involving a solid + acid.

Students may find it difficult to determine the crystallisation point. Advice such as 'Don't reduce the volume of the evaporating solution by more than half' might be helpful to them.

More quantitative work is included, requiring students to work out a theoretical yield to compare with the actual yield obtained.

Technician's notes

You will need the following resources for a class practical.

Be sure to check the latest safety notes on these resources before proceeding.

Part 1
burette and stand, 25 cm^3 pipette and bulb filler, conical flask, white tile, wash bottle containing distilled water
0.1M sodium hydroxide solution
0.15M hydrochloric acid, methyl orange
eye protection
Part 2
250 cm^3 beaker, measuring cylinder, spatula, glass rod, filter funnel and filter paper, evaporating basin
Bunsen burner, tripod and gauze, crystallizing dish, microscope slide
0.5M sulfuric acid
magnesium carbonate
access to an electronic balance accurate to 1 decimal place
eye protection

Answers

Page 214

1. A salt is formed when a replaceable hydrogen of an acid is replaced by a metal.
2. Sulfuric acid
3. Potassium chloride will be soluble in water (as are all potassium salts).
4. Calcium nitrate
5. Neutralisation is the reaction between and acid and an alkali or base to form a salt and water.
6. $H_+(aq)$
7. $OH^-(aq)$

Developing investigative skills, page 216

1. a) The pipette should be washed with distilled water and then a small amount of the potassium hydroxide solution (discard the solution used for washing).

b) The burette should be washed with distilled water and then a small amount of the sulfuric acid solution (discard the solution used for washing).

2. A pipette bulb (suction bulb) could be used to suck the potassium hydroxide solution into the pipette.

3. Using a funnel (and pouring below eye level by lifting the stand off the bench and down onto a stool).

4. Use a white tile or white card/paper under the conical flask. The methyl orange changes from yellow (in the potassium hydroxide solution) to pink (when too much sulfuric acid solution has been added).

5. Wash the conical flask with distilled water and discard the washings.

6. In the first experiment the student had no real idea how much acid would be needed. After the first experiment she knew approximately how much acid was needed and would be much more likely to get the colour change on adding just one drop.

7. 17.0 cm^3. The result from the first experiment is very different and likely to be the most inaccurate (as just explained).

8 Moles of H_2SO_4 added = 17.0/1000 × 0.1 = 0.0017

From the equation the ratio of H_2SO_4 : KOH is 1:2

Moles of KOH = 0.0034

Concentration of KOH = 1000/25 × 0.0034 = 0.136 mol/dm^3

C4a.3a Making a soluble salt (1) – using a titration method

A soluble salt can be made by reacting an acid with an alkali – a neutralisation reaction. To ensure that the quantities of acid and alkali exactly neutralise each other (the end-point), the reaction is first performed using an indicator (a titration reaction) and can then be repeated without the indicator. Typically, the acid is put in the burette and the alkali is measured in a pipette.

This activity is about the titration technique and requires a high degree of care to obtain an accurate result.

Apparatus

burette and stand, 25 cm^3 pipette and bulb filler, conical flask, white tile

dilute hydrochloric acid, dilute sodium hydroxide, wash bottle containing distilled water, methyl orange

eye protection

SAFETY INFORMATION

Dilute sodium hydroxide solution is an irritant.

Wear eye protection.

Method

1. Put on your eye protection.
2. Rinse the burette first with distilled water and then with the dilute hydrochloric acid. Discard the solution used.
3. Rinse the pipette first with distilled water and then with dilute sodium hydroxide solution. Discard the solution used.
4. Following the instructions from your teacher, use the pipette bulb filler to transfer exactly 25.00 cm^3 of dilute sodium hydroxide to a conical flask. Add 3 drops of methyl orange indicator.
5. Following the instructions from your teacher, fill the burette with dilute hydrochloric acid and make sure that there are no air bubbles near to the jet.
6. Take the initial volume of the acid in the burette. Add the acid to the sodium hydroxide solution in the conical flask, swirling the flask all the time, until the indicator changes colour. Take the final volume of the acid.
7. Wash out the conical flask with distilled water and then repeat the titration until two of the readings agree to within 0.10 cm^3. (It is likely that your first titration will not be very accurate – it is difficult to know how close you are to the end-point at your first attempt.)

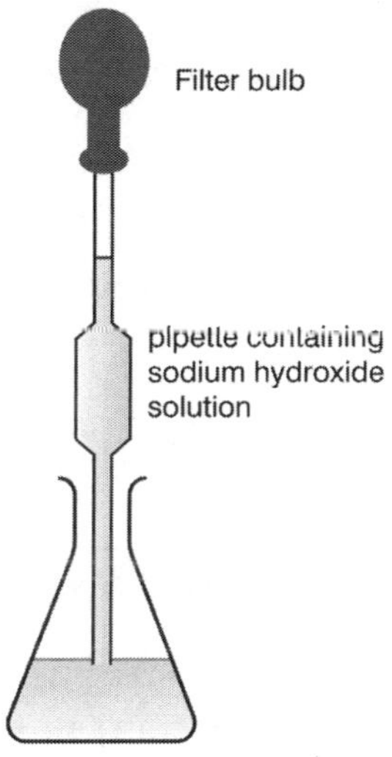

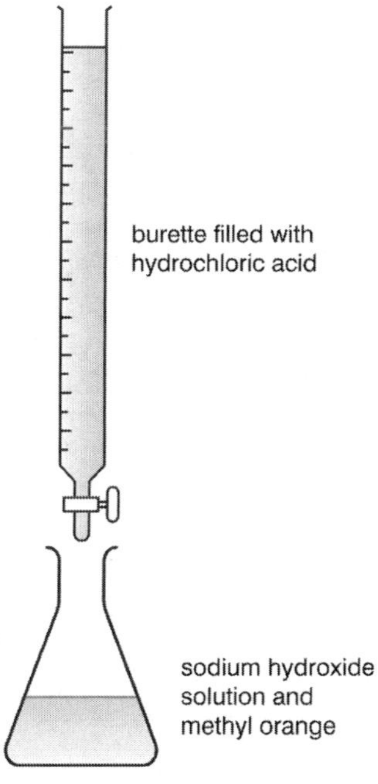

Note: The first result will give you a clear idea of the approximate amount of acid to add. In subsequent titrations you will be able to add the acid quickly until you are close to the end point and then add drop by drop. The indicator will change colour on the addition of 1 drop so you are aiming for accuracy of 1 drop!

Page 1 of 2

Make observations and measurements

Volume of sodium hydroxide solution used = 25.00 cm^3

Burette reading	1st titration	2nd titration	3rd titration	4th titration
2nd reading (cm^3)				
1st reading (cm^3)				
Difference (cm^3)				

Accurate volume of hydrochloric acid for neutralisation = ___________ cm^3 (x cm^3)

Analyse and interpret data

The titration technique can also be used to work out the concentration of either the acid or alkali used in the reaction. The calculation needed is presented step by step below.

Assume the concentration of the sodium hydroxide solution = 0.1M

Use your results to calculate the concentration of the hydrochloric acid. You know the accurate volume of solution used (shown as *x* above).

1. Write the equation for the reaction between hydrochloric acid and sodium hydroxide

____

2. How many moles of hydrochloric acid react with 1 mole of sodium hydroxide? ______________

3. The number of moles of sodium hydroxide used in 25.00 cm^3 of 0.1M solution = ______________

4. Therefore the number of moles of hydrochloric acid in x cm^3 of solution = ________________

5. Therefore the number of moles of hydrochloric acid in 1000 cm^3 = ____________________

6. The concentration of the hydrochloric acid solution = ______________M

C4a.3b Making a soluble salt (2)

Soluble salts can be made by adding an acid to a metal carbonate in the following reaction:

acid + metal carbonate → salt + carbon dioxide + water

In this activity you are going to make a sample of hydrated magnesium sulfate, $MgSO_4.7H_2O$

Apparatus

250 cm^3 beaker, measuring cylinder, spatula, glass rod, filter funnel and filter paper, evaporating basin, Bunsen burner, tripod and gauze, crystallizing dish, microscope slide

dilute sulfuric acid, magnesium carbonate, distilled water

eye protection

SAFETY INFORMATION

Dilute sulfuric acid is an irritant.

Wear eye protection.

Method

1. Put on your eye protection.

2. Put 50 cm^3 of sulfuric acid solution into a beaker. Add magnesium carbonate powder a little at a time until no further reaction occurs.

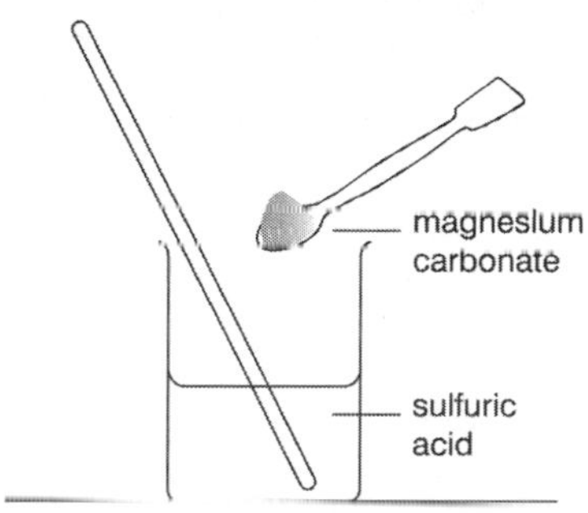

3. Filter the solution into an evaporating basin, leaving the unreacted magnesium carbonate on the filter paper.

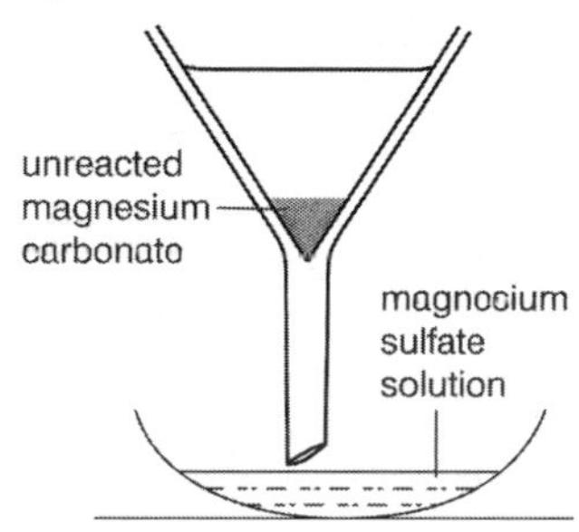

4. Heat the solution in the evaporating basin until it has reached crystallisation point (crystals form when the hot solution is added to a cold microscope slide). Stop heating at this point.

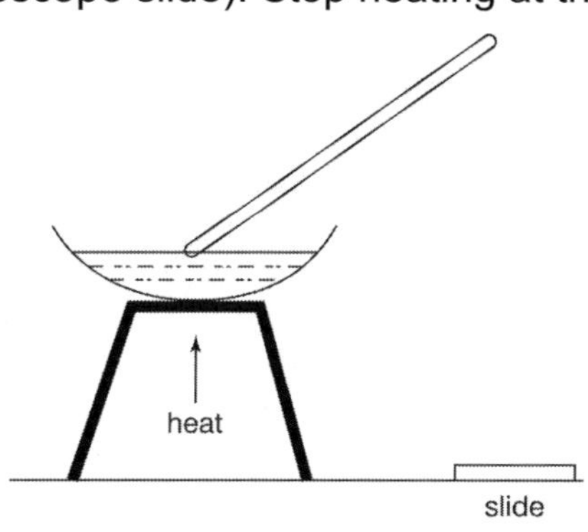

5. Pour the solution into a crystallising dish and leave to crystallise.

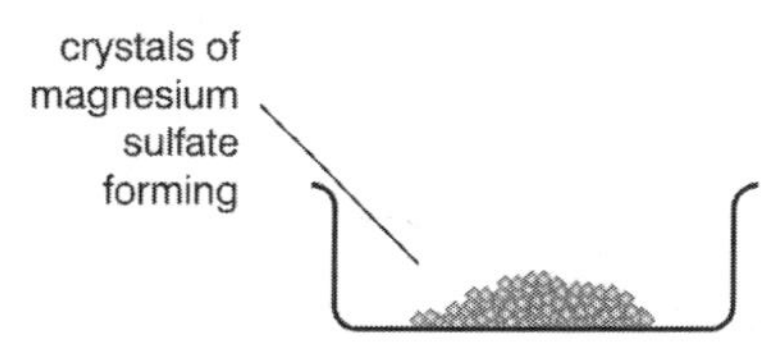

Page 1 of 2

Analyse and interpret data

1. What gas is produced in the reaction? ______________________________

2. What is the identification test for this gas? What would indicate a positive identification?

____

____

3. What does the term 'hydrated' mean?

____

4. Why isn't the evaporating basin containing the solution heated until all the liquid has evaporated?

____

____

5. A solution which will crystallise on cooling is said to be 'saturated'. What is a saturated solution?

____

____

6. The chemical equation for the reaction is:

$MgCO_3(s) + H_2SO_4(aq) + 7H_2O(l) \rightarrow MgSO_4.7H_2O(s) + CO_2(g) + H_2O(l)$

(A_r: H = 1; C = 12; O = 16; Mg = 24; S = 32)

In this reaction 50 cm^3 of 0.5M sulfuric acid was reacted with excess magnesium carbonate. Calculate the mass of magnesium sulfate crystals that should be produced (the theoretical yield).

____

____

____

____

____

7. The actual yield will be less than the theoretical yield. Explain the reason for this.

____

____

____

Activity C4a.4 Making insoluble salts

Learning objectives

- To introduce the preparation of insoluble salts using the precipitation method

Learning outcomes

- Be able to describe how to prepare an insoluble salt using a precipitation reaction

Resources

Student Book pages 214–217

Worksheet C4a.4a Making an insoluble salt

Files on CD–ROM: C4a.4a_worksheet, C4a.4a_tech_notes

Resources for class practical (see below)

Approach

Explain that the method for preparing an insoluble salt involves a precipitation reaction. Introduce the activity by referring to page 214 in the Student Book and then use Worksheet C4a.4a to enable students to carry out a number of different precipitation reactions.

The worksheet requires students to predict whether the combination of chemicals will produce an insoluble salt (thus revising the rules used to predict the solubility of a salt) and then to confirm or otherwise by trying the reaction. This is a good opportunity for students to practise writing chemical equations for the precipitation reactions.

Technician's notes

You will need the following resources for a class practical.

Be sure to check the latest safety notes on these resources before proceeding.

test tubes, test tube rack, droppers
0.1M lead(II) nitrate solution
0.1M solutions of sodium carbonate, sodium sulfate, silver nitrate
0.1M solutions of calcium nitrate, copper(II) sulfate, sodium chloride
0.1M solution of potassium sulfate
eye protection

Answers

Page 215

1. Precipitation is the formation of an insoluble salt as a result of a chemical reaction taking place in aqueous solution.

2. Filtration

3. Washing with cold water will remove traces of any remaining soluble salts.

4. a) lead(II) nitrate + sodium chloride → lead(II) chloride + sodium nitrate

b) $Pb(NO_3)_2(aq) + 2NaCl(aq) \rightarrow PbCl_2(s) + 2NaNO_3(aq)$

C4a.4a Making an insoluble salt

The following list will help you to predict whether a salt is soluble or insoluble:

- All common sodium and potassium salts are soluble.
- All nitrates are soluble.
- Common chlorides are soluble except silver chloride.
- Common sulfates are soluble, except those of barium and calcium.
- Common carbonates are insoluble, except those of sodium and potassium.

Apparatus

test tubes, test tube rack, droppers

solutions: sodium carbonate, sodium sulfate, silver nitrate, lead(II) nitrate, calcium nitrate, copper(II) sulfate, copper(II) nitrate, sodium chloride, potassium sulfate.

eye protection

SAFETY INFORMATION

Lead(II) nitrate is toxic.

Wear eye protection.

Method

1. Put on your eye protection.

2. Look at the combinations of solutions in the table below. Predict which combinations are likely to react and produce an insoluble salt and, if appropriate, give the name of the insoluble salt formed.

3. Mix the pairs of solutions in separate test tubes and record your observations in the table.

Solution A	Solution B	Do you predict the formation of an insoluble salt? If so, give the name of the insoluble salt	Observations on mixing the solutions
Sodium carbonate	Copper(II) sulphate		
Sodium sulfate	Copper(II) nitrate		
Silver nitrate	Sodium chloride		
Lead(II) nitrate	Sodium carbonate		
Calcium nitrate	Potassium sulfate		

Analyse and interpret data

1. How could a pure sample of an insoluble salt be separated from the remaining solutions?

____

____

2. Write chemical equations for the first two reactions occurring as you completed the table.

____

____

Activity C4a.5 Consolidation and summary

Learning objectives

- To review the learning points of the topic
- To test understanding through answering questions

Learning outcomes

- Be familiar with the knowledge and understanding summarised in the End of Topic Checklist
- Be able to apply this knowledge and understanding by answering End of Topic Questions

Resources

Student Book pages 217–221

Approach

Refer to the Science in Context feature which provides additional information on some acids and alkalis. You could use the Developing Investigative Skills feature on page 216 of the Student Book as a revision of the titration technique if you have not already used this as part of Activity C4a.3.

Ask students to study the End of Topic Checklist and then answer the End of Topic Questions.

Answers

Page 218

1. a) Magnesium carbonate + sulphuric acid → magnesium sulphate + carbon dioxide + water

   $MgCO_3 + H_2SO_4 \rightarrow MgSO_4 + CO_2 + H_2O$

   Note: Magnesium sulfate is hydrated and the correct formula is $MgSO_4.7H_2O$ (water could be added to both sides of the equation if this formula was required)

   b) Any risk assessment format could be used: ensure

   Hazard: Risk: Precautions are included

   All chemicals to be included (both reactants and products)

   All glassware

   Basic laboratory safety may be included but this could be listed or the lab rules referred to

   c) Equipment: beaker (100 cm^3/250 cm^3); measuring cylinder (250 cm^3); spatula; glass rod; filter paper and funnel; stand and clamp for filtering/tripod and gauze; Bunsen burner evaporating basin

   d) Instructions

   Complete and follow risk assessment during experiment.

   Measure (50 cm^3) dilute sulphuric acid into a beaker.

   Using a spatula add magnesium carbonate and stir the mixture. Continue adding the magnesium carbonate until fizzing stops and the magnesium carbonate is in excess.

   Filter the excess magnesium carbonate and collect the magnesium sulfate in an evaporating basin.

   Heat the solution in the evaporating basin until about half of the water has been removed. Leave the solution at room temperature and crystals of magnesium sulfate should form.

   e) The white powder is anhydrous magnesium sulfate – this is magnesium sulfate without the water of crystallisation. The presence of the water gives the crystalline structure; if this water is removed a powder is produced.

   f) In order to prepare a salt using an acid–base titration method both reactants need to be solutions. Magnesium carbonate is not soluble in water and therefore a standard solution can not be prepared.

   g) Anhydrous form formula $MgSO_4$. It can act as a drying agent as it readily absorbs water from the air. (It is hygroscopic.)

End of Topic Questions mark scheme

Question	Correct answer	Marks
1 a)	An indicator is a substance that will distinguish between an acid and an alkali.	1 mark
1 b)	The pH scale shows how strongly acidic or how strongly alkaline a substance is.	1 mark
1 c)	i) Weakly acidic ii) Weakly alkaline iii) Strongly alkaline	1 mark 1 mark 1 mark
2 a)	An acid is a substance that forms H^+ ions in water.	1 mark
2 b)	An alkali is a substance which forms OH^- ions in water.	1 mark

Question	Correct answer	Marks
2 c)	Neutralisation	1 mark
3 a)	A base	1 mark
3 b)	A salt	1 mark
3 c)	Soluble	1 mark
3 d)	The calcium oxide is added to warm dilute hydrochloric acid with stirring. When no more solid will react filter the mixture. Heat the filtrate in an evaporating basin until it is saturated/reaches crystallisation point (remove a sample onto a cold watch glass to see if it crystallises). Leave the solution to cool and crystallise.	1 mark 1 mark 1 mark 1 mark
3 e)	$CaO(s) + 2HCl(aq) \rightarrow CaCl_2(aq) + CO_2(g) + H_2O(l)$ (1 mark for balanced equation; 1 mark for state symbols)	2 marks
4 a)	Sulfuric acid	1 mark
4 b)	Barium chloride or barium nitrate	1 mark
4 c)	Mix solutions of dilute sulfuric acid and barium chloride/barium nitrate. Filter the suspension to isolate the solid. Wash the solid on the filter paper with distilled water and leave to dry/dry in a low temperature oven.	1 mark 1 mark 1 mark
4 d)	$BaCl_2(aq) + H_2SO_4(aq) \rightarrow BaSO_4(s) + 2HCl(aq)$ or $Ba(NO_3)_2(aq) + H_2SO_4(aq) \rightarrow BaSO_4(s) + 2HNO_3(aq)$ (1 mark for balanced equation; 1 mark for state symbols)	2 marks
5 a)	$K_2SO_4(aq) + 2H_2O(l)$ (1 mark for balanced equation; 1 mark for state symbols)	2 marks
5 b)	$MgCl_2(aq) + H_2O(l)$ (1 mark for balanced equation; 1 mark for state symbols)	2 marks
5 c)	$Ba(NO_3)_2(aq) + CO_2(g) + H_2O(l)$ (1 mark for balanced equation; 1 mark for state symbols)	2 marks
5 d)	$ZnCl_2(aq) + H_2(g)$ (1 mark for balanced equation; 1 mark for state symbols)	2 marks
5 e)	$ZnCO_3(s) + 2KCl(aq)$ (1 mark for balanced equation; 1 mark for state symbols)	2 marks
	Total:	34 marks

C4b Energetics

Introduction

This topic does not need much prior knowledge, although students must understand covalent bonding before they study the energy associated with bond breaking and bond forming in reactions.

There are many links to other topics: for example, energy change in displacement reactions links to Topic 2f *Reactivity series*; energy change in neutralisation reactions links to Topic 4a *Acids, alkalis and salts*; and the energy of combustion links to Topic 3a *Alkanes*, Topic 3c *Ethanol* and Topic 5b *Crude oil*. There are also links to Topic 4c *Rates of Reaction* (the activation energy of reactions) and to Topic 4d *Equilibria* (temperature affects the equilibrium of exothermic and endothermic reactions).

This topic starts with classifying reactions as exothermic and endothermic and then focuses on two types of calorimetric techniques – first using a polystyrene cup (for reactions such as displacement and neutralisation), and then a metal container (for combustion reactions). Energy level diagrams are then introduced, and the origin of these energy transfers is explained in terms of bond breaking and bond forming.

Finally, calculations involving average bond energies are introduced. There are good opportunities for students to develop their investigative skills, including opportunities for performing accurate quantitative work.

Links to other topics

Sections	Essential background knowledge	Useful links
1 Principles of chemistry	1e Chemical formulae and chemical equations 1g Covalent substances	
2 Chemistry of the elements		2f Reactivity series
3 Organic chemistry		3a Alkanes 3c Ethanol
4 Physical chemistry		4a Acids, alkalis and salts 4c Rates of reaction 4d Equilibria
5 Chemistry in society		5b Crude oil

Topic overview

C4b.1	**Orientation** Limited background knowledge is required for this topic, other than understanding of covalent bonding. This activity introduces the definitions of exothermic and endothermic reactions and then provides students with a number of practical examples of each type of reaction. Reactions include acid/alkali neutralisation reactions and displacement reactions. The physical change of dissolving potassium chloride in water is also introduced as work on 'cold packs' is developed in the next activity.

C4b.2	**Measuring energy transfers (1)** This activity is in two parts and introduces the calorimetric method using a polystyrene cup. Worksheets are available for each part. In the first part, the energy transfer in the displacement reaction between iron and copper(II) sulfate is calculated using the formula: energy transfer = mass of solution × specific heat capacity × temperature change The term 'enthalpy change' is introduced and the molar enthalpy change of this reaction is also calculated. The Developing Investigative Skills feature in the Student Book on measuring the molar enthalpy change in a neutralisation reaction could be used as part of this activity or as a revision in Activity C4b.5. The second part centres on an investigation into 'cold packs'. Students are required to design their own investigation to find out what quantities of potassium chloride and water must be mixed to provide a temperature change of –10 °C. This activity provides a good opportunity for students to apply what they have learnt about this simple calorimetric technique.
C4b.3	**Measuring energy transfers (2)** This activity involves the measurement of the enthalpy change of combustion of a fuel (ethanol) using a calorimetric method involving a metal can. An activity sheet is available to support the student investigative work.
C4b.4	**Energy level diagrams and bond breaking and bond forming** This activity introduces the concept of an energy level diagram and contrasts energy level diagrams for exothermic and endothermic reactions. The activation energy is introduced, and an energy level diagram is explained in terms of how it illustrates the difference in the energy required to break bonds and the energy released when bonds are formed. Finally, bond energy calculations are introduced.
C4b.5	**Consolidation and summary** This activity recaps on the ideas encountered in the topic. The Developing Investigative Skills feature on neutralisation in the Student Book could be used as a revision activity if it has not already been used as part of Activity C4b.2. Time can also be allocated to tackling the End of Topic Questions in the Student Book.

Activity C4b.1 Orientation

Learning objectives

- To revise understanding of covalent bonding
- To measure temperature changes in a range of reactions
- To classify reactions into those that give out energy and those that take energy in

Learning outcomes

- Know the nature of the bonds present in fuels such as the alkanes
- Know that chemical reactions in which heat energy is given out are described as exothermic
- Know that chemical reactions in which heat energy is taken in are described as endothermic

Common misconceptions

Dissolving is not a chemical reaction – dissolving potassium chloride in water is a physical change, not a chemical change.

Resources

Student Book pages 222–223

Worksheet C4b.1a Exothermic and endothermic reactions

Files on CD–ROM: C4b.1a_worksheet, C4b.1a_tech_notes

Resources for demonstration and class practical (see below)

Approach

Demonstrate the burning of magnesium ribbon in a Bunsen flame. Explain that the reaction generates a lot of heat energy – energy is transferred from the reaction to the surroundings. Such reactions are known as exothermic reactions. Do not look at the flame directly.

Ask the question 'Do all chemical reactions generate heat energy – are they all exothermic?' In fact, this is not the case. Introduce the terminology of an endothermic reaction – a reaction that absorbs energy from the surroundings.

Introduce the practical activity (Worksheet C4b.1a). This includes one reaction which results in a fall in the temperature of the surroundings (citric acid and sodium hydrogen carbonate) and a physical change (potassium chloride and water) which also absorbs energy from the surroundings. The other reactions (metal + acid; displacement of a metal ion; neutralisation) are exothermic.

Students may be familiar with metal + acid, displacement and neutralisation reactions. This is an opportunity to recap or to probe their knowledge. Explain that later in the topic they will study techniques for measuring the amount of heat energy released in exothermic and endothermic reactions, as well as the reasons behind these energy changes.

Technician's notes

You will need the following resources for a demonstration and a class practical.

Be sure to check the latest safety notes on these resources before proceeding.

Part 1
magnesium ribbon
tongs, Bunsen burner

Part 2
boiling tubes, boiling tube rack, thermometer, spatula, droppers
potassium chloride
magnesium ribbon
0.25M sulfuric acid, 0.5M citric acid, 0.25M sodium hydroxide
1M hydrochloric acid, 0.5M copper(II) sulfate
sodium hydrogen carbonate, iron filings
eye protection

C4b.1a Exothermic and endothermic reactions

Almost all chemical reactions involve a transfer of energy. In most reactions energy is transferred to the surroundings and the temperature of the surroundings increases – these reactions are EXOTHERMIC.

In a few reactions energy is absorbed from the surroundings and the temperature of the surroundings decreases – these reactions are ENDOTHERMIC.

Your task is to measure the temperature changes in a series of reactions in order to classify the reactions as exothermic or endothermic.

Apparatus

boiling tubes, boiling tube rack, thermometer, spatula, droppers

dilute hydrochloric acid, copper(II) sulfate solution, citric acid solution, dilute sulfuric acid, magnesium ribbon, iron filings, sodium hydrogen carbonate, sodium hydroxide solution, potassium chloride

eye protection

SAFETY INFORMATION

Magnesium ribbon is highly flammable.

Sulfuric acid, citric acid and sodium hydroxide are all irritants.

Wear eye protection.

Method

1. Put on your eye protection.
2. Put about a 2 cm depth of substance 1 into a boiling tube. Record the temperature of the liquid or solution.
3. Add either 1 heaped spatula measure or, if a solution, about a 2 cm depth of substance 2.
4. Stir the reaction mixture and take the highest or lowest temperature reached.

Page 1 of 2

Make observations and measurements

Substance 1	Substance 2	Initial temperature (°C)	Final temperature (°C)	Temperature change (°C)	Exothermic or endothermic
Dilute hydrochloric acid	Magnesium ribbon				
Copper(II) sulfate solution	Iron filings				
Citric acid solution	Sodium hydrogen carbonate				
Dilute sulfuric acid	Sodium hydroxide solution				
Water	Potassium chloride				

Analyse and interpret data

1. In one of the combinations of substances the temperature change is not the result of a chemical change or chemical reaction. Which one? Explain why this combination of substances could not be described as a chemical reaction.

____

____

____

2. Add ‘exothermic’ or ‘endothermic’, as appropriate, to the last column in the table.

Page 2 of 2

Activity C4b.2 Measuring energy transfers (1)

Learning objectives

- To calculate energy changes in a reaction using a polystyrene cup as a simple calorimeter
- To calculate energy changes in terms of the molar quantities given in the equation

Learning outcomes

- Know how a polystyrene cup can be used to carry out simple calorimetric experiments to determine energy changes
- Understand how to use the equation:
  energy transferred = mass of water (g) × 4.2 (J/g/°C) × rise in temperature (°C)
- Understand that heat energy is called enthalpy
- Know that the molar enthalpy change (ΔH) is the heat energy change when the molar quantities shown in the chemical equation react together

Resources

Student Book pages 222–225, 227

Worksheet C4b.2a Measuring energy transfers (1)

Worksheet C4b.2b Designing a 'cold pack'

Files on CD–ROM: C4b.2a_worksheet, C4b.2b_worksheet, C4b.2a_tech_notes, C4b.2b_tech_notes

Resources for class practical (see below)

Approach

This activity is in two parts, each using the calorimetric technique with a polystyrene beaker. Explain that the following formula is used to measure the energy change in a chemical reaction:

energy transfer = mass of solution × specific heat capacity × temperature change

Take care when explaining the units used for each quantity. To give the energy in kJ, the mass is expressed in kg, the specific heat capacity in kJ/kg/°C and the temperature change in °C. For laboratory investigations, using the mass (g), specific heat capacity (J/g/°C) and rise in temperature (°C) gives the energy in J, which can then be converted to kJ. This is more convenient.

1. Measuring energy transfers

This involves calculating the energy change in the displacement reaction between iron and copper(II) sulfate solution. Details are given on Worksheet C4b.2a.

The insulating properties of the polystyrene justify its use. A lid could be used to further reduce heat transfer.

Ask students to use the formula above to calculate the energy change in the reaction they perform (that is, with specific quantities). Explain that for comparison it is better to have the energy change per mole, or using the molar quantities expressed in the equation. Consequently, students calculate the molar energy change. Explain the term 'enthalpy' and introduce the term 'molar enthalpy change'.

The Developing Investigative Skills feature on page 227 of the Student Book focuses on the molar enthalpy change of neutralisation. You could introduce this as consolidation at this point or left to Activity C4b.5 at the end of the topic.

2. Designing a 'cold pack'

This part is intended to provide some open-ended investigational work with the students required to plan and design their own approach to identifying the quantities of potassium chloride and water in a 'cold pack' that will produce a drop in temperature of 10 °C.

Explain that cold packs can be used to limit the impact of sports injuries. Some manufacturers use potassium chloride and water in these products. Worksheet C4b.2b provides some structure for the investigation.

Encourage students to consider what a systematic approach might require – not just to use trial and error. One approach is to repeat the experiment 3 or 4 times using different quantities of potassium chloride with a fixed volume of water, then plot results on a graph and extrapolate to identify the quantities required for the temperature drop required.

Students are required to evaluate the procedure they used and to suggest how it could be improved.

Technician's notes

You will need the following resources for a class practical.

Be sure to check the latest safety notes on these resources before proceeding.

Part 1
polystyrene cup, thermometer accurate to 0.1 °C, measuring cylinder, spatula, access to electronic balance
0.5M copper(II) sulfate solution, iron filings
eye protection
Part 2
polystyrene cup, thermometer accurate to 0.1 °C, measuring cylinder, spatula, glass rod, access to electronic balance
potassium chloride
eye protection

Answers

Page 224

1. A reaction that releases heat energy to the surroundings.
2. A reaction that absorbs energy from the surroundings.
3. Polystyrene is a very good insulator and so very little energy is transferred to the surroundings.
4. Energy change = 100 × 4.2 × 6 = 2520 J

Page 225

1. A high proportion of the energy released is transferred to the surrounding air.
2. Energy = 30 × 4.2 × 15 = 1890 J for 0.1 g corn puffs. Energy for 1 g corn puffs = 1890 × 10 = 18 900 J
3. Energy = 200 × 4.2 × 14 = 11 760 J. Energy per mole = 11 760 × 46/0.5 = 1 081 920 J (1082 kJ)

Developing investigative skills, page 227

1. Measuring cylinders or burettes.
2. Wear protective gloves and goggles. Mop up any spillages quickly with a damp cloth.
3. Putting a lid on the polystyrene cup.
4. Enthalpy change = 100 × 4.2 × 7 = 2940 J
5. Molar enthalpy change = 2940 × 1000/50 (the solutions each contained 0.05 moles)
   = 58 800 J/mol or 58.8 kJ/mol $\Delta H = -58.8$ kJ/mol

6. As the same volume of each solution was used, the ‘starting temperature’ could have been taken as the average of the two. So, for example, if the sodium hydroxide was at 20 °C and the hydrochloric acid at 18 °C, the starting temperature would have been 19 °C.

7. Possible errors were: loss of heat energy during the reaction (likely to be the greatest); and heat energy absorbed by the polystyrene.

C4b.2a Measuring energy transfers (1)

Energy transfers in chemical reactions can be measured using a container called a calorimeter. In this activity a polystyrene cup acts as a very effective calorimeter. You are going to measure the energy transferred in the displacement reaction between iron and copper(II) sulfate.

$$Fe(s) + CuSO_4(aq) \rightarrow FeSO_4(aq) + Cu(s)$$

Apparatus

polystyrene cup, thermometer accurate to 0.1 °C, measuring cylinder, spatula, access to an electronic balance

copper(II) sulfate solution, iron filings

eye protection

SAFETY INFORMATION
Wear eye protection.

Method

1. Put on your eye protection.
2. Measure 20 cm^3 of copper(II) sulfate solution into a polystyrene cup.
3. Record the temperature of the solution in the table below.
4. Add approximately 1 g of iron filings to the solution, a spatula at a time and stir with the thermometer.
5. Record the highest temperature reached in the table below.

Make observations and measurements

Volume of copper(II) sulfate solution used = _______ cm^3

Initial temperature (°C)	Final temperature (°C)	Temperature change (°C)

Analyse and interpret data

The energy transferred in the reaction can be measured using the following equation:

energy transferred = mass of solution × specific heat capacity × change in temperature

You can assume that:

- 1 cm^3 of copper(II) sulfate solution has a mass of 1 g (its density is 1 g/cm^3)
- The specific heat capacity of copper(II) sulfate solution is 4.2 J/g/°C.

Page 1 of 2

The calculation can be carried out in 2 parts.

1. Calculate the energy transferred in the reaction:

2. In the reaction 20 cm^3 of 0.5M copper(II) sulfate was used, which is equivalent to 0.01 mole. The iron was in excess. Calculate the energy transferred if molar quantities had been used in the reaction.

Evaluate data and methods

1. What is the main advantage of using a polystyrene cup in this experiment?

2. How could the accuracy of the experiment be improved?

3. What does the sentence 'The iron was in excess' mean?

C4b.2b Designing a 'cold pack'

Commercially produced 'cold packs' are designed to produce a cooling effect, which can be useful for treating sports injuries. It is likely that cold packs contain two substances which are originally in separate compartments within the cold pack.

When you bend the pack, the barrier between the two substances breaks and they mix or react together. Whether the cooling effect is a consequence of a chemical reaction, or simply a physical process such as dissolving, may depend on the brand of the cold pack.

In this investigation you will investigate the cooling effect produced when potassium chloride is added to water.

Devise and plan investigation

Your task is to find out:

What quantities of potassium chloride and water are needed to produce a drop in temperature of 10 °C.? (The cold pack must hold between 50 cm^3 and 100 cm^3 of liquid.)

Some things to think about:

- Which method or technique will I use?
- What equipment is suitable and will give the accuracy I need?
- What quantities of potassium chloride and water should I use?
- What risk assessments should I make? What are my safety precautions?
- What will I actually do (my method)?

Method

Write your method below:

Page 1 of 2

Make observations and measurements

Draw a table to record your results.

Analyse and interpret data

Show how you used your results to decide on the quantities of potassium chloride and water to use in the cold pack.

Quantities to use in cold pack: potassium chloride __________ g water ____________ cm^3

Evaluate data and methods

1. Overall did the method chosen give accurate results?

____

____

2. If you were to do the investigation again, what aspects would you change? Give reasons for changes you would make.

____

____

____

____

____

Activity C4b.3 Measuring energy transfers (2)

Learning objectives

- To measure energy changes during the combustion of a fuel
- To calculate energy changes in terms of the molar quantities given in the equation

Learning outcomes

- Know how a metal calorimeter can be used to measure the energy change when a fuel is burnt
- Understand how to use the equation:
  energy transferred = mass of water (g) × 4.2 (J/g/°C) × rise in temperature (°C)
- Understand that heat energy is called enthalpy
- Know that the molar enthalpy change (ΔH) is the heat energy change when the molar quantities shown in the chemical equation react together
- Understand how to calculate the molar enthalpy change from the results of an experiment and information about molar quantities

Resources

Student Book pages 222–225, 227

Worksheet C4b.3a Measuring energy transfers (2)

Files on CD–ROM: C4b.3a_worksheet, C4b.3a_tech_notes

Resources for class practical (see below)

Approach

This activity uses a different calorimetric technique – one that will provide an estimate of the energy released when a fuel (ethanol) burns. This provides a link to Topic 3c *Ethanol*.

A metal container is used as the calorimeter because the metal is a good conductor of heat and will transfer the energy from the flame to the water. The technique involves considerable energy loss but can provide comparative results for a number of different fuels.

The technique is explained on Worksheet C4b.3a. A worked example of a typical calculation is provided on page 224 of the Student Book.

Warn students to be extremely careful when handling the spirit burner (often the tops of the burners are loose fitting and the ethanol can readily escape) – if it was knocked over when burning, it could cause a substantial fire.

The *Analyse and interpret data* section of the worksheet requires students to calculate the molar enthalpy change. The importance of giving the sign for a molar enthalpy change (negative for an exothermic reaction and positive for endothermic reaction) is covered in the next activity, which focuses on energy level diagrams.

Technician's notes

You will need the following resources for a class practical.

Be sure to check the latest safety notes on these resources before proceeding.

tin or copper can, thermometer, heat resistant boards to use as draught shields, measuring cylinder, access to electronic balance
spirit burner and lid containing ethanol
eye protection

Answers

Page 224

1. A reaction that releases heat energy to the surroundings.
2. A reaction that absorbs energy from the surroundings.
3. Polystyrene is a very good insulator and so very little energy is transferred to the surroundings.
4. Energy change = 100 × 4.2 × 6 = 2520 J

Page 225

1. A high proportion of the energy released is transferred to the surrounding air.
2. Energy = 30 × 4.2 × 15 = 1890 J for 0.1 g corn puffs. Energy for 1 g corn puffs = 1890 × 10 = 18 900 J
3. Energy = 200 × 4.2 × 14 = 11 760 J. Energy per mole = 11 760 × 46/0.5 = 1 081 920 J (1082 kJ)

Developing investigative skills, page 227

1. Measuring cylinders or burettes.
2. Wear protective gloves and goggles. Mop up any spillages quickly with a damp cloth.
3. Putting a lid on the polystyrene cup.
4. Enthalpy change = 100 × 4.2 × 7 = 2940 J
5. Molar enthalpy change = 2940 × 1000/50 (the solutions each contained 0.05 moles)

   = 58 800 J/mol or 58.8 kJ/mol

   ΔH = –58.8 kJ/mol

6. As the same volume of each solution was used, the 'starting temperature' could have been taken as the average of the two. So, for example, if the sodium hydroxide was at 20 °C and the hydrochloric acid at 18 °C, the starting temperature would have been 19 °C.

7. Possible errors were: loss of heat energy during the reaction (likely to be the greatest); and heat energy absorbed by the polystyrene.

C4b.3a Measuring energy transfers (2)

A calorimetric technique can be used to find out the energy transferred when a fuel burns. The apparatus is shown in the diagram below. The calorimeter is a tin can.

In this activity you will be able to find out how much energy is transferred when ethanol (C_2H_5OH) burns.

Apparatus

tin can, thermometer, draught screens

clamp and stand

access to an electronic balance, measuring cylinder

ethanol in a spirit burner

eye protection

SAFETY INFORMATION

Ethanol is highly flammable.

Wear eye protection.

Method

1. Put on your eye protection.
2. Put 100cm^3 (100 g) of water into the calorimeter.
3. Assemble the apparatus as shown in the diagram and take the temperature of the water.
4. Measure the mass of the spirit burner and ethanol.
5. Light the spirit burner and heat the water in the calorimeter, stirring all the time with the thermometer, until the temperature has risen by about 20 °C.
6. Take the exact final temperature of the water and immediately reweigh the spirit burner.

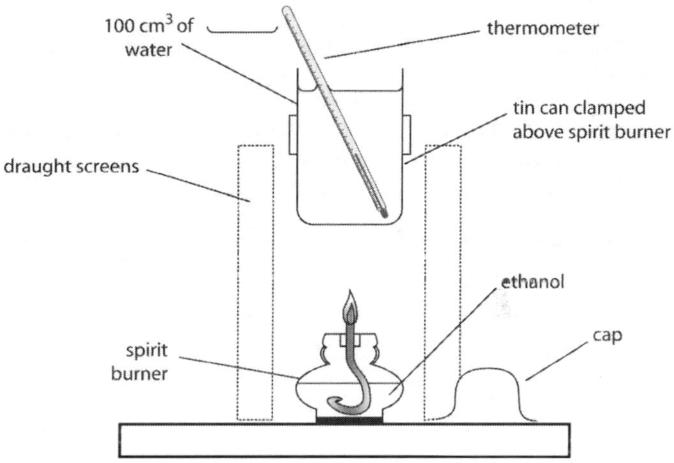

Make observations and measurements

Record your results in the table.

Mass of water being heated	g
Initial temperature of the water	°C
Final temperature of the water	°C
Rise in temperature of the water	°C
Mass of burner + ethanol at the start of the experiment	g
Mass of burner + ethanol at the end of the experiment	g
Mass of ethanol burned	g

Page 1 of 2

Analyse and interpret data

The energy transferred in the reaction can be measured using the following equation:

energy transferred = mass of water × specific heat capacity × change in temperature

Remember:

- 1 cm^3 of water has a mass of 1 g (its density is 1 g/cm^3)
- The specific heat capacity of water solution is 4.2 J/g/°C.

1. Calculate the energy transferred in your reaction.

2. What energy would have been transferred if 1 g of ethanol had been burned?

3. What energy would have been released if 1 mole of ethanol had been burned (the molar enthalpy change)?

(A_r: H = 1, C = 12, O = 16)

Evaluate data and methods

1. Why was a tin can used in this experiment rather than a glass beaker?

2. Your value for the molar enthalpy change is likely to be much lower than the value quoted in data books. List the sources of errors in your experiment.

Activity C4b.4 Energy level diagrams and bond breaking and bond forming

Learning objectives

- To introduce energy level diagrams
- To highlight the differences in energy level diagrams for exothermic and endothermic reactions
- To show the significance of bond breaking and bond making in the determination of the molar enthalpy change (ΔH) for a reaction
- To use bond energy values to calculate molar enthalpy changes in reactions

Learning outcomes

- Understand that ΔH is used to represent the molar enthalpy change for a reaction
- Know that ΔH for an exothermic reaction is negative and that ΔH for an endothermic reaction is positive
- Understand how energy level diagrams can be used to show the difference between exothermic and endothermic reactions
- Understand that the breaking of bonds is endothermic and that the making of bonds is exothermic
- Be able to use average bond energies to calculate the enthalpy change in a simple chemical reaction

Resources

Student Book pages 226, 228–230

Approach

Energy profiles and energy level diagrams are introduced on page 226 of the Student Book, followed by an explanation in terms of bond breaking and bond forming on pages 228–230. Then introduce bond energy calculations. You may use the following sequence:

- Explain the use of simple energy level diagrams to represent exothermic and endothermic reactions; emphasise the importance of using negative (exothermic) and positive (endothermic) signs for ΔH values.
- Explain the extension of these ideas to an energy level profile, which includes the concept of the activation energy, the minimum amount of energy needed to initiate a reaction between reacting particles.
- Explain the origin of the energy changes in exothermic and endothermic reactions in terms of the energy needed to break bonds (an endothermic process) and the energy released on forming bonds (an exothermic process). Using page 228 of the Student Book, illustrate this in terms of the alkane, propane. Refer to the trend in the molar enthalpy of combustion in the alkane series. This provides an opportunity to review knowledge of the alkane homologous series if Topic 3ab *Alkanes* has been studied.
- Demonstrate how average bond energies can be used to calculate molar enthalpy changes in reactions. Make use of the table of average bond energies and the worked example in the Student Book. Use the questions in the Student Book to consolidate understanding and to practise using average bond energies in simple calculations.

Answers

Page 226

1. Endothermic
2. The activation energy

Page 230

1. The sign indicates whether the reaction is exothermic (negative sign) or endothermic (positive sign).
2. Energy is needed to break bonds.
3. In an endothermic reaction more energy is needed to break bonds than is recovered on forming bonds.
4. The units are kJ/mol.
5. $2H_2(g) + O_2(g) \rightarrow 2H_2O(l)$

   Energy needed to break bonds = (2 × 436) + 498 = +1370 kJ

   Energy released on forming bonds = 4 × 464 = –1856 kJ

   Energy change = (+1370) – (1856) = –486 kJ

Activity C4b.5 Consolidation and summary

Learning objectives

- To review the learning points of the topic
- To test understanding through answering questions

Learning outcomes

- Be familiar with the knowledge and understanding summarised in the End of Topic Checklist
- Be able to apply this knowledge and understanding by answering the End of Topic Questions

Resources

Student Book pages 231–235

Approach

It is worth spending some time to consolidate the learning in the topic, particularly the quantitative aspects, which some students may find challenging.

Provide examples of calculations for the two calorimetric methods covered in the topic, as well as for bond energy calculations. The Developing Investigative Skills feature in the Student Book on measuring the molar enthalpy of neutralisation could be used here if it has not been covered earlier.

Consider the End of Topic Checklist as a basis for revising ideas and give students opportunities to answer the End of Topic Questions.

The Science in Context feature on endothermic reactions on page 231 of the Student Book can also be used.

Answers

Page 232

1. a) $2H_2(g) + O_2(g) \rightarrow 2H_2O(l)$

b) In an exothermic reaction energy is transferred to the surroundings and the temperature rises.

c) Exothermic reaction

d)

Bonds broken (energy absorbed)		Bonds formed (energy released)	
O=O	498	O-H × 4	464 × 4
H-H × 2	436 × 2 = 872		
Total	1370		1856

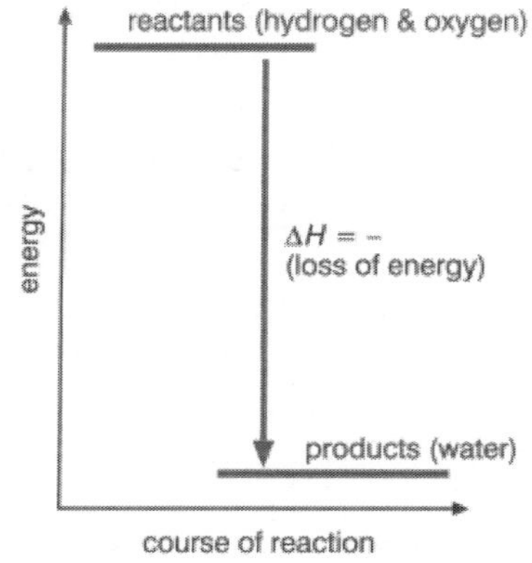

Overall energy change 1856 – 1370 = 486 kJ

Equation $2H_2(g) + O_2(g) \rightarrow 2H_2O(l)$

Therefore overall energy change is 486/2 = 243 kJ/mol of hydrogen burned.

2. At present it is expensive to manufacture hydrogen by using electrolysis of water. (Although obtaining hydrogen from electrolysis using wind or solar generated electricity is now being considered.)

Hydrogen also requires much more storage volume than the volume of petrol that produces the equivalent energy.

End of Topic Questions mark scheme

Question	Correct answer	Marks
1 a)	Energy = mass × specific heat capacity × temperature change	1 mark
	Energy = 40 × 4.2 × 32	1 mark
	Energy = 5376 J	1 mark
1 b)	Energy per g of magnesium = 5376 × 1.0/0.2 = 26 880 J (26.88 kJ)	1 mark
1 c)	Energy per mole of magnesium = 26.88 × 24 J	1 mark
	Energy per mole of magnesium = 645.12 kJ/mole (allow 645 kJ)	1 mark
2 a)	A metal container	1 mark
	Metal is a good conductor of heat.	1 mark
2 b)	A considerable amount of energy is lost to the environment.	1 mark
	Not all the energy is transferred to the water.	1 mark
2 c)	Surround the fuel and container with heat resistant mats.	1 mark
	To exclude drafts.	1 mark
3 a)	The reaction is exothermic.	1 mark
3 b)	The energy released on forming new bonds is greater than the energy used to break the bonds.	1 mark
4	The energy needed to break the H–H and Cl–Cl bonds	1 mark
	Is less than the energy released on forming H–Cl bonds	1 mark
5 a)	Energy needed to break bonds = 4E(C–H) + 2E(O=O) = (4 × 413) + (2 × 498) = 2648 kJ	1 mark
	Energy released on forming bonds = 2E(C=O) + 4E(O–H) = (2 × 745) + (4 × 464) = 3346 kJ	1 mark
	Energy change = 2648 – 3346 = -698 kJ/mole	1 mark
	(Must include the negative sign)	
5 b)	The reaction is exothermic.	1 mark
	Total:	20 marks

C4c Rates of reaction

Introduction

This topic extends the idea of the particulate nature of matter to develop a simple collision theory that explains why certain factors can change the rate of a reaction. A study of catalysts provides links to the preparation of oxygen in Topic 2d *Oxygen and oxides* and the preparation of hydrogen in Topic 2e *Hydrogen and water*. Activation energy provides a link with Topic 4b *Energetics* and the link with Topic 4d *Equilibria* provides a basis for consideration of the industrial processes in Topic 5d *The industrial manufacture of chemicals*.

The topic lends itself to a range of different investigative activities. Once you have introduced the experimental techniques for measuring the rate of a reaction, the factors – surface area, concentration, temperature and catalysts – can be investigated. The concept of an energy profile is used to explain the effect of a catalyst on a reaction.

Links to other sections

Sections	Essential background knowledge	Useful links
1 Principles of chemistry	1a States of matter 1b Atoms 1e Chemical formulae and chemical equations	1g Covalent substances
2 Chemistry of the elements		2d Oxygen and oxides 2e Hydrogen and water
4 Physical chemistry		4b Energetics 4d Equilibria
5 Chemistry in society		5d The industrial manufacture of chemicals

Topic overview

C4c.1	**Orientation** This activity brings together a number of different ideas: • A recap on the students' current knowledge and understanding of the particulate nature of matter • A quick demonstration of reactions that occur at different speeds • An explanation of the difference between rate, speed and time • An introduction to a simple collision theory.
C4c.2	**Monitoring the rate of a reaction** This activity involves practical work based on the reaction of marble chips with dilute hydrochloric acid and magnesium and hydrochloric acid. The rate of the two reactions is monitored either by measuring the loss in mass of the reaction mixture as the carbon dioxide escapes or by measuring the volume of hydrogen gas produced. The collision theory can be used to explain why the rate of reaction changes.

C4c.3	**Surface area of solids** This activity involves practical work using apparatus similar to that used in Activity C4c.2. The reaction of 'large' pieces of marble chip with hydrochloric acid can be compared with the same reaction involving 'small' pieces of marble chip. The collision theory explains why the rate of reaction is affected by surface area of solids.
C4c.4	**Concentration of solutions** This activity involves practical work using the reaction between sodium thiosulfate solution and dilute hydrochloric acid. The concentration of the sodium thiosulfate solution is changed and the time taken for a pencilled cross to be obscured by the precipitated sulfur is measured. The collision theory can be used to explain why the rate of reaction is affected by the concentration of solutions.
C4c.5	**Temperature** This involves practical work, either using the sodium thiosulfate and acid reaction used in Activity C4c.4 or one of the other reactions encountered in the topic. The collision theory explains why the rate of reaction is affected by temperature.
C4c.6	**Use of a catalyst** This involves practical work focused on identifying a good catalyst for the decomposition of hydrogen peroxide solution. In addition, the effect of changing the amount of the catalyst on the rate of the reaction is demonstrated. The effect of a catalyst on the reaction is explained in terms of an energy level profile.
C4c.7	**Consolidation and summary** This activity allows for all the learning in the topic to be consolidated and for key ideas to be applied through tackling the End of topic questions in the Student Book.

Activity C4c.1 Orientation

Learning objectives

- To revise the particle arrangements and movement in the three states of matter
- To demonstrate that reactions occur at different rates
- To introduce a simple collision theory

Learning outcomes

- Be able to explain how particles are arranged and move in solids, liquids and gases
- Know the difference between time, speed and rate
- Know in terms of a collision theory what must happen for two particles to react together

Common misconceptions

Rate/speed and time are the same.

All collisions between reacting particles result in a reaction.

Resources

Student Book pages 236–237

Worksheet C4c.1a The particulate nature of matter

Files on CD–ROM: C4c.1a_worksheet, C4c.1a_tech_notes

Resources for demonstration (see below)

Approach

1. Recap on students' current knowledge and understanding of the particulate nature of matter.

Introduce the topic using page 236 of the Student Book, concentrating on the introduction, knowledge check and learning objectives. Start with the card sort activity on Worksheet C4c.1a where small groups of students arrange the cards into 'true', 'false' and 'don't know'.

Join each group with another to discuss any differences in their groupings. Take feedback and summarise the key differences between the arrangement and movement of the particles in solids, liquids and gases.

2. Quick demonstration of reactions that occur at different speeds.

Demonstrate a few reactions to show the different speeds at which the reactions take place:

a) Instant	Lead(II)nitrate solution and potassium iodide solution
b) Rapid once started	Burning magnesium ribbon in air (oxygen)
c) Quite slow	Sodium thiosulfate solution and dilute hydrochloric acid
d) Very slow	A shiny steel nail and most air (no observable change)

Ask the question: 'Why do some reactions happen very quickly and others are very slow?'

3. Explanation of the difference between speed, rate and time.

Ask students what 'rate' is. Refer to a car: its speed in miles/hour (kilometres/hour) is in fact the rate of travel of the car. Refer to the table in the Student Book on speed, rate and time.

4. Introduction to a simple collision theory.

Explain the collision theory model that scientists use to explain the different rates at which reactions occur. Emphasise that there are two requirements:

- The reacting particles must collide.
- There must be sufficient energy involved in the collision to break chemical bonds. These collisions are called effective collisions.

Technician's notes

You will need the following resources for a demonstration.

Be sure to check the latest safety notes on these resources before proceeding.

0.1M lead(II) nitrate solution (50 cm^3)
0.1M potassium iodide solution (50 cm^3)
magnesium ribbon (10 cm)
0.2 M sodium thiosulfate solution (50 cm^3)
2M hydrochloric acid (10 cm^3)
4 beakers (100 cm^3)
crucible, tongs
Bunsen burner
eye protection

Answers

Worksheet C4c.1a

1. False 2. True 3. True 4. True 5. True 6. True 7. True
8. False 9. True 10. True 11. True 12. False

Notes:

1. The particles in a solid are vibrating and not stationary.

11. When ice melts the particles move closer together (due to the nature of the hydrogen bonds in the ice) – this is not the general occurrence when a solid melts.

12. The process is more accurately known as diffusion and not dispersion.

C4c.1a The particulate nature of matter

Cut out the cards and then sort them into piles of 'True', 'False' or 'Not sure'.

1. In a solid the particles are not moving.	2. The particles in a gas are very far apart.	3. In a liquid the particles will be constantly bumping into each other.
4. The particles in a gas are moving at high speed.	5. The particles in a solid are held together by various types of forces.	6. Solid A has a higher melting point than solid B so the forces holding the particles together in A are stronger than those in B.
7. As a liquid is heated the particles move faster.	8. As a solid is heated the particles stay still.	9. The forces between gas particles are very small.
10. When a liquid evaporates particles escape from the surface of the liquid.	11. When a solid melts and forms a liquid the particles usually move further apart.	12. When gas particles mix together this is known as dispersion.

Activity C4c.2 Monitoring the rate of reaction

Learning objectives

- To demonstrate that the rate of a reaction usually decreases as the reaction takes place
- To show some of the experimental methods that can be used to measure the changing rate of a reaction
- To show how experimental results can be used to work out the rate of a reaction at a particular time

Learning outcomes

- Be able to demonstrate and describe techniques showing how the rate of a reaction can be monitored by either collecting the gas produced or measuring the loss in mass as the gas escapes
- Know how to make observations and measurements in reactions where the change in rate is being measured
- Understand how to analyse and interpret data to draw conclusions about the rate of a reaction
- Be able to evaluate data and methods in the experiments used
- Be able to explain why rate is affected by concentration using a simple collision theory model

Common misconceptions

Reactions proceed at a constant rate.

Resources

Student Book pages 238–240

Worksheet C4c.2a Monitoring the rate of a reaction

Files on CD–ROM: C4c.2a_worksheet, C4c.2a_tech_notes

Resources for demonstration and class practical (see below)

Approach

Your choice of approach will depend on the apparatus available. You will need an electronic balance reading to two decimal places to accurately measure the loss in mass in the reaction between marble chips and dilute hydrochloric acid.

As for measuring the volume of gas produced in a reaction, a gas syringe is ideal but an inverted measuring cylinder in a trough of water is an acceptable alternative. There are opportunities to practise of all the investigative skills.

One approach is to:

1. Demonstrate how the rate of the reaction between marble chips and hydrochloric acid can be monitored by measuring the loss in mass as the carbon dioxide escapes. The apparatus and method are given in the Student Book page 239. Record the class results and ask students to show them graphically. Discuss the different regions of the graph and link the steepness of the gradient to the rate of the reaction.

2. Use a class practical to explore the reaction between dilute hydrochloric acid and magnesium ribbon by measuring the volume of hydrogen produced. Full details are provided on Worksheet C4c.2a. **CAUTION: Make sure there are no naked flames which could ignite the magnesium or the hydrogen produced.**

3. You may use the graph from the class practical to reinforce the interpretation of the changing rate of reaction as a reaction proceeds. Use a simple collision theory to explain why the rate decreases – as the reactant particles are used up in the reaction, there will be fewer effective collisions.

Technician's notes

You will need the following resources for a demonstration and class practical.

Be sure to check the latest safety notes on these resources before proceeding.

1. Demonstration of the reaction between marble chips and dilute hydrochloric acid.

5 g of marble chips (an excess)
30 cm^3 of 0.25M hydrochloric acid
measuring cylinder
electronic balance (reading to 2 decimal places)
conical flask and cotton wool plug
stopwatch
eye protection

2. Class practical of the reaction between magnesium and dilute hydrochloric acid.

8 cm magnesium ribbon
25 cm^3 1M hydrochloric acid
conical flask, bung and delivery tube
gas syringe or 100 cm^3 measuring cylinder and trough of water
stand and clamp
eye protection

Answers

Page 238

1. a) The particles must collide.
   b) There must be sufficient energy in the collision (to break bonds).
2. An effective collision is one which results in a chemical reaction between the colliding particles.
3. It is an energy barrier. Only collisions which have enough energy to overcome this barrier will lead to a reaction.

Page 240

1. A gas syringe will accurately measure the volume of gas produced.
2. No gas is being produced – the reaction hasn't started or it is finished.
3. The quicker reaction will have the steeper gradient.

C4c.2a Monitoring the rate of a reaction

What happens to the rate of a reaction as it proceeds? In this investigation you will be able to monitor the rate of the reaction between magnesium and hydrochloric acid.

$Mg(s) + 2HCl(aq) \rightarrow MgCl_2(aq) + H_2(g)$

Apparatus

8 cm of magnesium ribbon

25 cm^3 of dilute hydrochloric acid

conical flask, bung and delivery tube

gas syringe or 100cm^3 measuring cylinder and trough of water

stand and clamp

stopwatch

eye protection

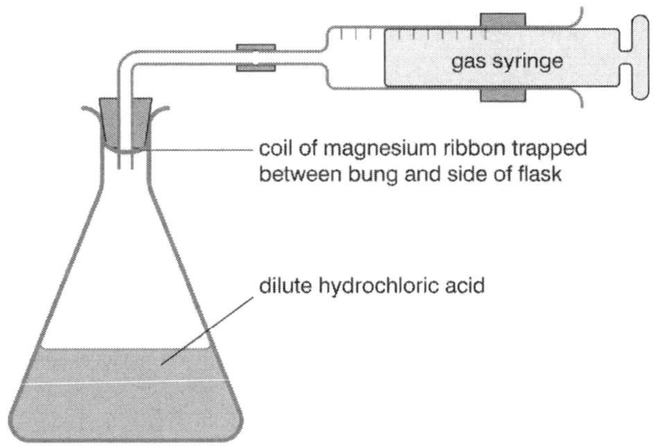

SAFETY INFORMATION

Hydrogen gas is extremely flammable.

There should be no flames in the room.

Wear eye protection.

Method

1. Put on your eye protection.
2. Prepare a table of results as shown on the right.
3. Carefully transfer 25 cm^3 hydrochloric acid into the conical flask.
4. Make a coil with the magnesium ribbon. Carefully trap it between the bung and the side of the flask (as shown in the diagram above).
5. Twist the bung until the magnesium ribbon falls into the acid, start the stopwatch immediately. Make sure the bung fits tightly.
6. Measure the total volume of hydrogen collected in the syringe every 10 seconds.

Time (s)	Total volume of hydrogen (cm^3)
0	
10	
20	
30	
40	
50	
60	
70	
80	
90	

Analyse and interpret data

1. Plot a graph of total volume of gas against time. Draw a smooth curve through the points.
2. After how many seconds did the reaction finish?
3. How does the shape of the graph show that the reaction has finished?
4. What does the shape of the graph tell you about how the rate of reaction changes?
5. Use the idea of collisions between particles to explain why the rate changes during the reaction.

Activity C4c.3 Surface area of solids

Learning objectives

- To show that the rate of a reaction involving a solid and a liquid/solution depends on the surface area of the solid
- To show how the effect of surface area on rate can be measured experimentally
- To explain the effect of surface area on the rate of a reaction using a simple collision model

Learning outcomes

- Know how to make observations and measurements in reactions where the change in rate is being measured
- Understand how to analyse and interpret data to draw conclusions about the rate of a reaction
- Be able to evaluate data and methods in the experiments used
- Be able to explain why rate is affected by surface area using a simple collision theory model

Common misconceptions

Students may need convincing that the same mass of smaller particles will have a greater surface area than bigger particles. The model of a sliced loaf of bread can be used help explain this.

Resources

Student Book pages 240–245

Worksheet C4c.3a Reaction rate and surface area

Files on CD–ROM: C4c.3a_worksheet, C4c.3a_tech_notes

Resources for demonstration and class practical (see below)

Approach

Whichever practical approach you use, students will be familiar with the apparatus from the previous activity. The reaction will be slower than that of magnesium ribbon and dilute hydrochloric acid, so you have to decide for how long students should take readings.

It will be sufficient to compare the initial rates or early stages of the two reactions rather than waiting for the reaction to go to completion. Ideally, students should not be taking readings for more than 5 or 6 minutes for each experiment. (About 90 cm^3 of carbon dioxide should be produced at room temperature. If you want a quicker reaction you can increase the concentration of the acid, but make sure the syringe is disconnected once the volume reaches 100 cm^3.)

The results can be shown graphically and the gradients of the two curves compared in the early stages of the reactions. You may use the collision theory to explain why the reaction has a greater rate with smaller particles – more marble surface and calcium carbonate 'particles' (ions) will be exposed to the hydrochloric acid particles (H^+ ions), so there will be more effective collisions in a given time.

Technician's notes

The reaction between marble chips and dilute hydrochloric acid can be monitored by measuring the loss in mass or the volume of carbon dioxide produced as the reaction proceeds. The availability of accurate electronic balances (accurate to 0.01 g) will determine whether this method is used, either as a demonstration or class practical. The method involving the collection of gas might be easier to resource. The marble chips can be categorised roughly into 'large' and 'small'.

You will need the following resources for a demonstration and a class practical.

Be sure to check the latest safety notes on these resources before proceeding.

Demonstration

5 g of 'large' marble chips (an excess)
5 g of 'small' marble chips (an excess)
2 × 30 cm^3 of 0.25M hydrochloric acid (allow 100 cm^3)
measuring cylinder
electronic balance (reading to 2 decimal places)
conical flask and cotton wool plug
stopwatch
eye protection

Class practical, per group:

5 g 'large' marble chips (an excess)
5 g 'small' marble chips (an excess)
2 × 30 cm^3 0.25M hydrochloric acid (allow 100 cm^3)
conical flask, bung and delivery tube
gas syringe or 100 cm^3 measuring cylinder and trough of water
stand and clamp
eye protection

Answers

Page 244

1. The units of concentration for solutions are mol/dm^3.

2. The particles are more closely packed together and so there will be more (effective) collisions per second.

3. Increasing temperature means the particles: have more (kinetic) energy; more of the collisions will have energy greater than or equal to the activation energy; there will be more effective/successful collisions per second are moving faster and so will collide more often and with more energy in the collisions.

C4c.3a Reaction rate and surface area

What effect does the surface area of a solid have on the rate at which it reacts? In this investigation you are going to study the reaction between marble chips (calcium carbonate) and hydrochloric acid.

$$CaCO_3(s) + 2HCl(aq) \rightarrow CaCl_2(aq) + CO_2(g) + H_2O(l)$$

Apparatus

5 g ‘large’ marble chips (an excess)

5 g ‘small’ marble chips (an excess)

30 cm^3 0.25M hydrochloric acid (for each experiment)

conical flask, bung and delivery tube

gas syringe or 100 cm^3 measuring cylinder and trough of water

stand and clamp

stopwatch

eye protection

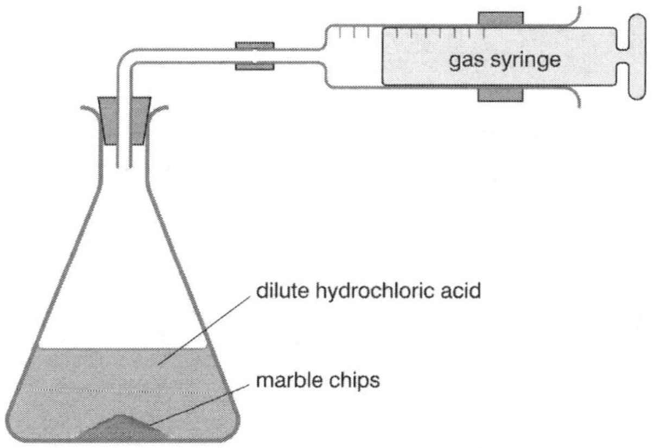

SAFETY INFORMATION

Wear eye protection.

Method

1. Put on your eye protection.

2. Add 30 cm^3 of the hydrochloric acid solution to 5 g of ‘small’ marble chips in a conical flask. Immediately connect to the gas syringe and start the stop watch.

3. Take the total volume of gas produced every 30 seconds for about 5 minutes, recording your results in a table like the one shown on the right.

4. Carefully wash out the conical flask and then repeat the experiment with 5 g of ‘large’ marble chips. Record your results in the table.

Time (s)	Total volume of gas (‘small’ marble chips) (cm^3)	Total volume of gas (‘large’ marble chips) (cm^3)
0	0	0

Analyse and interpret data

1. Plot the results to the two experiments on graph paper using the same axes.

2. What do the shapes of the two curves tell you about the differences in the rates of the reactions?

3. Use the idea of collisions between reacting particles to explain the differences in reaction rate.

Activity C4c.4 Concentration of solutions

Learning objectives

- To show that the rate of a reaction depends on the concentration of solutions
- To show how the 'initial' rate of a reaction can be measured with different concentrations of reactants
- To explain the effect of concentration on the rate of a reaction using a simple collision model

Learning outcomes

- Be able to demonstrate and describe a technique for comparing the rates of a reaction with different concentrations of solutions
- Know how to make observations and measurements in reactions where the change in rate is being measured
- Understand how to analyse and interpret data to draw conclusions about the rate of a reaction
- Be able to evaluate data and methods in the experiment used
- Be able to explain why rate is affected by concentration using a simple collision theory model

Common misconceptions

In the sodium thiosulfate and acid experiment, the time taken to produce a fixed amount of sulfur (enough to obliterate the pencilled mark) is used as a measure of the rate of the different reactions. This may, therefore, cause some confusion between time and rate.

Resources

Student Book pages 240–245

Worksheet C4c.4a Reaction rate and concentration

Files on CD–ROM: C4c.4a_worksheet, C4c.4a_tech_notes

Resources for class practical (see below)

Approach

The Developing Investigative Skills box pages 242–243 of the Student Book relates directly to the experiment used in this activity.

The equation for the reaction is:

$Na_2S_2O_3(aq) + 2HCl(aq) \rightarrow 2NaCl(aq) + S(s) + SO_2(g) + H_2O(l)$

The sulfur produced as a precipitate in the reaction provides a means of easily comparing the rates of the reaction as the concentration of sodium thiosulfate solution is changed.

CAUTION: Sulphur dioxide produced in the reaction. In solution it is a LOW HAZARD but the gas is TOXIC. Students with asthma should be warned of this hazard, and good ventilation is needed.

The concentration of the sodium thiosulfate is changed by diluting the standard solution with water. The concentration of the hydrochloric acid is kept constant. To be able to easily relate volumes to concentrations the total volume of the reaction mixture should be always the same.

A pencil mark is often better than one produced in ink as the latter will 'run' if it gets wet. The results should produce a smooth curve (plotting volume of sodium thiosulfate against 1/time should produce a straight line graph).

Explain the results in terms of the collision theory: the more particles in a certain volume of solution the more effective collisions there will be. Do not use temperatures above 50 ˚C.

Technician's notes

You will need the following resources for a class practical.

Be sure to check the latest safety notes on these resources before proceeding.

40 cm^3 2M hydrochloric acid
200 cm^3 sodium thiosulfate solution (40 g per litre)
100 cm^3 measuring cylinder (sodium thiosulfate solution)
100 cm^3 measuring cylinder (water)
10 or 25 cm^3 measuring cylinder (acid)
conical flask
stop watch
eye protection

Answers

Developing investigative skills, page 242

1. To ensure that it was a fair test – the depth of liquid above the cross would always be the same.

2. The reaction starts as soon as the hydrochloric acid is added – during the stirring and positioning of the flask onto the pencil cross the reaction is taking place.

3. A burette or measuring cylinder. (The burette will be more accurate).

4. The greater the volume of sodium thiosulfate (that is, the greater the concentration of the solution) the greater the rate of the reaction.

5. 45 s (range 42 to 48 s).

6. a) Deciding when the pencilled cross is no longer visible (it disappears quite slowly, particularly for the more dilute solutions)

   b) Starting the stop clock consistently at the same point in mixing the solutions. Losing liquid while swirling or stirring is another possibility.

Page 244

1. The units of concentration for solutions are mol/dm^3.

2. The particles are more closely packed together and so there will be more (effective) collisions per second.

3. Increasing temperature means the particles: have more (kinetic) energy; more of the collisions will have energy greater than or equal to the activation energy; there will be more effective/successful collisions per second are moving faster and so will collide more often and with more energy in the collisions.

C4c.4a Reaction rate and concentration

Using more concentrated chemicals costs more money. Is it worth the extra cost? What are the advantages and disadvantages? From this investigation you will find out what effect changing concentration has on the rate of a reaction.

The reaction of sodium thiosulfate solution with hydrochloric acid produces sulfur as one of the products. The sulfur forms a pale yellow suspension in the solution. This fact can be used to investigate the rate of the reaction.

Apparatus

40 cm^3 2M hydrochloric acid

200 cm^3 sodium thiosulfate solution (40 g per litre)

100 cm^3 measuring cylinder (for sodium thiosulfate solution)

100 cm^3 measuring cylinder (for water)

10 or 25 cm^3 measuring cylinder (for acid)

conical flask, stop watch

eye protection

SAFETY INFORMATION

Sulfur dioxide gas produced in small quantities in this experiment is toxic.

Avoid inhaling it.

Wear eye protection.

Method

1. Put on your eye protection.
2. Mix the sodium thiosulfate and water in the conical flask using the volumes shown in the table below.

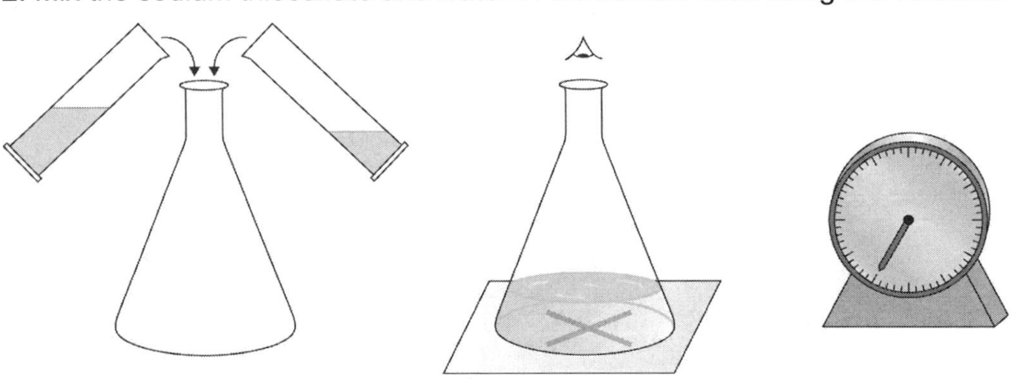

3. Add the acid to the flask, swirl the content and start the stopwatch.
4. Place the flask on a pencilled mark on some white paper.
5. Record the time when the cross can no longer be seen.

Page 1 of 2

Volume of sodium thiosulfate solution (cm^3)	50	40	30	20	10
Volume of water (cm^3)	0	10	20	30	40
Volume of hydrochloric acid (cm^3)	5	5	5	5	5
Time for mark to disappear (s)					

Analyse and interpret data

1. Why was the total volume of sodium thiosulfate and water always kept at 50 cm^3?
2. Plot a graph of your results and draw a smooth curve through the points.
3. Plot a graph of volume of sodium thiosulfate against 1/time (a measure of rate).

Evaluate data and methods

1. List the main sources of error in the experiment.

Activity C4c.5 Temperature

Learning objectives

- To show that the rate of a reaction depends on temperature
- To show how the rate of a reaction can be measured at different temperatures
- To explain the effect of temperature on the rate of a reaction using a simple collision model

Learning outcomes

- Be able to devise and plan an investigation to show how temperature affects the rate of a reaction
- Know how to make observations and measurements in reactions where rate is being measured
- Understand how to analyse and interpret data to draw conclusions about the rate of a reaction
- Be able to evaluate data and methods in the experiment used
- Be able to explain why rate is affected by temperature using a simple collision theory model

Common misconceptions

The dramatic impact of temperature may come as a surprise to students – a 10 °C rise in temperature can double the rate of a reaction.

Resources

Student Book pages 240–245

Worksheet C4c.5a Reaction rate and temperature

Files on CD–ROM: C4c.5a_worksheet, C4c.5a_tech_notes

Resources for class practical (see below)

Approach

The 'disappearing mark' experiment lends itself to an investigation of temperature.

CAUTION: Sulphur dioxide produced in the reaction. In solution it is a LOW HAZARD but the gas is TOXIC. Asthmatics should be warned of this hazard.

This activity is a very good opportunity for students to devise and plan their own investigation. They will be familiar with the approach used in the previous activity, but will have to decide how to adapt it to allow an investigation of temperature. It is important to allow students to produce a plan without too much prompting.

Choosing the concentration of sodium thiosulfate solution to use in all experiments is a key aspect. If they choose the fastest reaction from the previous activity, they will find the reaction is too rapid to measure accurately. Choosing the slowest will be much more successful (10 cm^3 of sodium thiosulfate solution and 40 cm^3 of water).

The other challenges are to work out what to heat to achieve temperatures above room temperature (ideally the thiosulfate/water mixture) and when to record the temperature (after the hydrochloric acid has been added). The maximum temperature used should be 50 °C.

Once students have completed their investigation, their results can be explained in terms of the collision theory. A crucial point is that increasing temperature will not only ensure more collisions; it will significantly increase the energy of the collisions.

Technician's notes

You will need the following resources for a class practical.

Be sure to check the latest safety notes on these resources before proceeding.

40 cm^3 2M hydrochloric acid
100 cm^3 sodium thiosulfate solution (40 g per litre)
100 cm^3 measuring cylinder (sodium thiosulfate solution)
100 cm^3 measuring cylinder (water)
10 or 25 cm^3 measuring cylinder (acid)
conical flask
stop watch
eye protection

Answers

Page 244

1. The units of concentration for solutions are mol/dm^3.

2. The particles are more closely packed together and so there will be more (effective) collisions per second.

3. Increasing temperature means the particles: have more (kinetic) energy; more of the collisions will have energy greater than or equal to the activation energy; there will be more effective/successful collisions per second are moving faster and so will collide more often and with more energy in the collisions.

C4c.5a Reaction rate and temperature

Rusting occurs slowly at room temperature but is much quicker at higher temperatures. How much difference does temperature actually make to the rate of a reaction?

Your task is to design and plan an investigation to find out the effect of temperature on the rate of reaction between sodium thiosulfate and hydrochloric acid.

Devise and plan investigation

How are you going to change the temperature?

How are you going to control the other variables? (The maximum temperature you should use is 50 °C.)

What is the best mixture of sodium thiosulfate and water to use?

What range of temperatures would be safe and sensible?

What are the hazards (what is your assessment of risk)?

How can the risks be reduced?

SAFETY INFORMATION

Make sure you have your plan checked before you start to carry it out.

Demonstrate and describe techniques

1. Carry out your approach.

Make observations and measurements

1. Record your results.

Analyse and interpret data

1. What do your results show? Can you explain them using a simple collision theory?

Evaluate data and methods

1. Where any of your results inaccurate? If so, how did you recognise them and what were the likely causes?

2. How could your approach have been improved?

Activity C4c.6 Use of a catalyst

Learning objectives

- To show that catalysts can increase the rate of a reaction
- To demonstrate the specificity of catalysts
- To show how manganese(IV) oxide catalyses the decomposition of hydrogen peroxide
- To explain how a catalyst increases the rate of a reaction

Learning outcomes

- Know how to make observations and measurements in reactions where rate is being measured
- Understand how to analyse and interpret data to draw conclusions about the rate of a reaction
- Be able to evaluate data and methods in the experiment used
- Be able to explain why rate is affected by the presence of a catalyst

Common misconceptions

Catalysts always increase the rate of a reaction. Some catalysts or inhibitors reduce the rate of a reaction.

Resources

Student Book pages 246–247

Worksheet C4c.6a Catalysts

Files on CD–ROM: 4c.6a_worksheet, 4c.6a_tech_notes

Resources for class practical and demonstration (see below)

Approach

Students should now be familiar with the experimental methods used to investigate the rate of various reactions. It may be that they do not need any further practise or experience. A suggested approach is as follows:

1. Students do a quick test tube investigation of possible catalysts for the decomposition of hydrogen peroxide (see Worksheet C4c.6a). They add spatula measures of the solids to approximately 5 cm^3 of hydrogen peroxide in a series of test tubes. They observe the tubes to see if effervescence occurs (oxygen formation) and, if it does, at what rate.

Note: Manganese (IV) oxide and copper (II) oxide are HARMFUL as the fine powder can be inhaled.

2. A teacher demonstration is used to reinforce the extent of the catalytic effect of manganese(IV) oxide on hydrogen peroxide – possibly comparing the effect of changing the amount of catalyst involved.

The way a catalyst increases the rate of a reaction – by offering a reaction pathway with a lower activation energy needs to be stressed. The specificity of catalysts (and enzymes for that matter) is also worthy of attention. Enzymes are mentioned briefly in the Student Book but some research on enzymes (and possibly enzyme inhibitors) would be a useful extension activity.

Technician's notes

You will need the following resources for a class practical and demonstration.

Be sure to check the latest safety notes on these resources before proceeding.

Class practical

40 cm^3 of 20 volume hydrogen peroxide solution
a range of metal oxides (examples overleaf)

copper(II) oxide
iron(III) oxide
manganese(IV) oxide
calcium oxide
magnesium oxide
aluminium oxide
test tubes and test tube rack
teat pipette
spatula
eye protection

Demonstration

100 cm^3 of 20 volume hydrogen peroxide solution
manganese(IV) oxide
conical flask with delivery tube to gas syringe
100 cm^3 measuring cylinder
spatula
eye protection

Answers

Page 247

1. A catalyst is a substance that changes the rate of a chemical reaction.
2. A biological catalyst is called an enzyme.
3. Changes in pressure affect reactions involving gases.

C4c.6a Catalysts

Catalysts are substances that change the rate of a chemical reaction. In most cases they increase the rate of the reaction. Catalysts are often specific to one reaction; in other reactions they have little or no effect.

Your task is to find the best catalyst for increasing the rate of decomposition of hydrogen peroxide into water and oxygen:

$2H_2O_2(aq) \rightarrow 2H_2O(l) + O_2(g)$

The reaction is very slow at room temperature without a catalyst.

Apparatus

40 cm^3 of 20 volume hydrogen peroxide solution

copper(II) oxide, iron(III) oxide, manganese(IV) oxide, calcium oxide, magnesium oxide, aluminium oxide

test tubes and test tube rack

teat pipette

spatula

eye protection

SAFETY INFORMATION

Hydrogen peroxide solution is an irritant.

Copper(II) oxide and manganese(IV) oxide are both harmful.

Wear eye protection.

Method

1. Put on your eye protection.
2. Add a spatula measure of each solid to separate samples of 5 cm^3 of hydrogen peroxide solution in test tubes.
3. Look for signs of oxygen being produced.

Results

1. Rank your solids in order from the most effective to the least effective at breaking down the hydrogen peroxide into oxygen and water.

MOST EFFECTIVE ______________________

____

____

____

____

LEAST EFFECTIVE ______________________

Activity C4c.7 Consolidation and summary

Learning objectives

- To revise the key learning points of the topic
- To test understanding through answering questions

Learning outcomes

- Be familiar with the knowledge and understanding summarised in the End of Topic Checklist
- Be able to apply this knowledge and understanding by answering the End of Topic Questions

Resources

Student Book pages 248–251

Approach

The revision should be active, involving tackling questions and discussing ideas. There are questions at the end of the topic as well as examination-style questions at the end of the Physical Chemistry section. Examination questions often require knowledge of practical techniques and experiments, the analysis and interpretation of data and evaluation of data and methods.

Answers

End of Topic Questions mark scheme

Question	Correct answer	Marks
1	There must be enough energy in the collision. To exceed the activation energy/break bonds	1 mark 1 mark
2 a)	The activation energy is an energy barrier, it is the minimum energy required for a reaction to take place.	1 mark
2 b)	Reaction A is likely to have the greater rate of reaction at a particular temperature. It has a lower activation energy.	1 mark 1 mark
3 a)	Carbon dioxide	1 mark
3 b)	$CaCO_3(s) + 2HCl(aq) \rightarrow CaCl_2(aq) + CO_2(g) + H_2O(l)$ (1 mark for balanced equation; 1 mark for state symbols)	2 marks
3 c)	Marks for the graph: Suitable scale for axes Points plotted correctly Smooth curve drawn	 1 mark 1 mark 1 mark
3 d)	i) 20 cm^3 ii) 16 cm^3 iii) 13 cm^3 iv) 0 cm^3	1 mark 1 mark 1 mark 1 mark

Question	Correct answer	Marks
3 e)	The rate of reaction is greatest at the beginning of the reaction. The rate decreases as the reaction proceeds and is zero after 70 seconds.	1 mark 1 mark
3 f)	The concentration of the reactants is greatest at the beginning of the reaction and so there will be more effective collisions. As the reactants are used up there will be fewer effective collisions and so the rate of the reaction will decrease.	1 mark 1 mark
3 g)	Marks for the graph: The line should start from the origin. The curve should be steeper/have a greater gradient and not cross the other curve. The curve should reach the same plateau but sooner.	 1 mark 1 mark 1 mark
3 h)	The question does not state which reactant is in excess so there are some possibilities. Marks for the graph: The line should start from the origin. The curve should be less steep/have a smaller gradient and not cross the first curve. The curve should reach a plateau below or at the same level as the original plateau but after a longer time.	 1 mark 1 mark 1 mark
4	As the particles will be moving quicker there will be more collisions per second. There will also be more energy available in the collision to overcome the activation energy/break bonds	1 mark 1 mark
5 a)	The catalyst provides an alternative pathway for the reaction The pathway has a lower activation energy	1 mark 1 mark
5 b)	Manganese(IV) oxide (allow alternatives)	1 mark
5 c)	Phosphoric acid	1 mark
6	Increasing pressure will increase the rate of reaction The gas particles will move closer together so there will be more effective collisions per second	1 mark 1 mark
	Total:	33 marks

C4d Equilibria

Introduction

This short topic extends the basic ideas on chemical equations introduced in Topic 1e *Chemical formulae and chemical equations*. Reversible reactions are introduced through the action of heat on copper(II) sulfate crystals and ammonium chloride. This is followed by an activity focusing on how changing the conditions of a reaction affects the position of equilibrium.

There are some important links to other topics: the use of anhydrous copper(II) sulfate to identify the presence of water in Topic 2e *Hydrogen and water*; hydrated salts in Topic 4a *Acids, alkalis and salts*; exothermic and endothermic reactions in Topic 4b *Energetics*; catalysts in Topic 4c *Rates of reaction*; and to important industrial processes involving reversible reactions in Topic 3c *Ethanol* and Topic 5d *The industrial manufacture of chemicals*.

Students have the opportunity to develop their investigative skills in the *Make observations and measurements* category.

Links to other topics

Sections	**Essential background knowledge**	**Useful links**
Principles of chemistry	1e Chemical formulae and chemical equations	
Chemistry of the elements		2e Hydrogen and water
Organic chemistry		3c Ethanol
Physical chemistry		4a Acids, alkalis and salts 4b Energetics 4c Rates of reaction
Chemistry in society		5d The industrial manufacture of chemicals

Topic overview

C4d.1	**Orientation** This activity introduces reversible reactions by focusing on the action of heat on hydrated copper(II) sulfate and ammonium chloride. In the case of the copper(II) sulfate, water is added to the residue (anhydrous copper(II) sulfate) to illustrate the reversibility of the reaction.
C4d.2	**Changing the position of equilibrium** This activity introduces the expression 'dynamic equilibrium' and then considers how changing conditions can have an impact on the equilibrium position in reversible reactions. It also refers to the 'balancing act' between maximising the rate of a reaction and obtaining the best equilibrium position.
C4d.3	**Consolidation and summary** This activity recaps on the key ideas encountered in the topic. Time can be allocated to studying the End of Topic Checklist and then answering the End of Topic Questions in the Student Book.

Activity C4d.1 Orientation

Learning objectives

- To introduce the concept of a reversible reaction
- To study the actions of heat on copper(II) sulfate crystals and ammonium chloride

Learning outcomes

- Know that a reversible reaction is one in which reactants form products and products can reform reactants
- Know that some reactions, such as heating copper(II) sulfate or ammonium chloride are reversible
- Know the symbol used in a chemical equation to show a reversible reaction
- Be able to describe the changes that occur when copper(II) sulfate crystals are heated and then when water is added
- Be able to describe the changes that occur when ammonium chloride is heated

Resources

Student Book pages 252–254

Worksheet C4d.1a Reversible reactions

Files on CD–ROM: C4d.1a_worksheet, C4d.1a_tech_notes

Resources for a class practical (see below)

Approach

Introduce reversible reactions as reactions that can go 'both ways': that is, from reactants to products and products back to reactants. Students may or may not be familiar with the action of heat on copper(II) sulfate and the terms 'hydrated' and 'anhydrous'.

Explain that the action of heat on copper(II) sulfate and ammonium chloride can be used to demonstrate two reversible reactions.

Use Worksheet C4d.1a to support the practical work and encourage students to produce full observations which include the appearance of the substances before, during and after the heating.

The hydrated copper(II) sulfate will lose its water of crystallisation and form anhydrous copper(II) sulfate. Adding water to the anhydrous copper(II) sulfate will reverse the change.

In the case of the ammonium chloride, decomposition occurs to form hydrogen chloride and ammonia which will then recombine in the cooler parts of the tube.

Technician's notes

You will need the following resources for a class practical.

Be sure to check the latest safety notes on these resources before proceeding.

test tubes, test tube holder, test tube rack, spatula, Bunsen burner
copper(II) sulfate crystals, ammonium chloride (no more than 5 g)
eye protection

C4d.1a Reversible reactions

Some chemical reactions are reversible, which means that reactants can be changed into products but the products can change back into the reactants. You are going to observe two reversible reactions.

Apparatus

Test tubes, test tube holder, test tube rack, spatula, Bunsen burner

Copper(II) sulfate crystals (hydrated copper(II) sulfate), ammonium chloride

eye protection

SAFETY INFORMATION
Copper(II) sulfate crystals and ammonium chloride are harmful.
Wear eye protection.

Method

1. Put on your eye protection.

2. Gently heat about a 1 cm depth of copper(II) sulfate crystals in a test tube. When no further change occurs, allow the tube to cool for a few minutes. Then add a few drops of water to the solid in the tube. Record your observations in the table below. Stop heating when it all turns white.

3. Use a low flame to heat about a 1 cm depth of ammonium chloride in a test tube. Leave the tube to cool. Record your observations in the table below.

Make observations and measurements

Reaction	Observations
1. Copper(II) sulfate crystals a) Heating copper(II) sulfate crystals b) Cool and then add water to the solid produced	
2. Ammonium chloride a) Heating ammonium chloride b) Cool	

Analyse and interpret data

1. What does the term 'hydrated' mean?

____

2. What is the name of the solid formed when copper(II) sulfate crystals are heated?

____

3. Write equations for the two reactions:

____

____

Activity C4d.2 Changing the position of equilibrium

Learning objectives

- To introduce the concept of a dynamic equilibrium
- To consider the effects of changing the conditions of a reaction on the position of the equilibrium

Learning outcomes

- Understand and be able to explain the concept of a dynamic equilibrium
- Be able to predict the effects of the following changes on a reversible or equilibrium reaction:
  changing the concentration of reactants or products
  changing the pressure
  changing the temperature
- Know that a catalyst affects the rates of forwards and backward reactions to the same extent and so does not change the position of equilibrium

Common misconceptions

Students will often confuse rate and equilibria.

Resources

Student Book pages 254–258

Approach

Explain the concept of a dynamic equilibrium using analogies such as walking up an escalator that is moving down or running on a treadmill. Contrast this with a static equilibrium (such as a chair standing on a level floor), which is a more common experience in everyday life.

Explain that reversible reactions, or reactions in this state of dynamic equilibrium, are problematic if you are trying to make a product and don't want it to revert to the reactant as soon as it has formed. Refer to the section on pages 254–255 of the Student Book on 'Changing the position of equilibrium'. Concentrate on the impact of changing the pressure and temperature on a reaction in equilibrium.

Link exothermic and endothermic reactions to Topic 4b *Energetics*. It is important to emphasise that these considerations are different to those relating to changing the rate of a reaction. In particular, stress that catalysts have no impact on equilibrium position – a catalyst will have equal impact on forward and reverse reactions.

Refer to the Science in Context feature 'Equilibria – it's certainly a balancing act!' which addresses the issue of getting a suitable rate of reaction at the same time as getting a suitable (percentage) yield (equilibrium position).

Answers

Page 254

1. The copper(II) sulfate crystals can be converted into anhydrous copper(II) sulfate by heating and removing the water in the crystals. When the water is added back to anhydrous copper(II) sulfate the copper(II) sulfate crystals are re-formed.

2. The concentration of each of the products and reactants remains constant unless a change is made to the reaction.

3. Reactants are constantly being converted into products and products are constantly being converted back into reactants. The rates of these two reactions are the same.

Page 257

1. a) Low temperature

   b) High pressure

   c) Using a catalyst has no impact on the position of equilibrium.

2. a) The catalyst increases the rate of the reaction (but has no affect on the position of equilibrium).

   b) The rate of the reaction would be very low.

Page 258

1. a) Magnesium + hydrochloric acid → magnesium chloride + hydrogen

   $Mg(s) + 2HCl(aq) \rightarrow MgCl_2(aq) + H_2(g)$

   b) i)

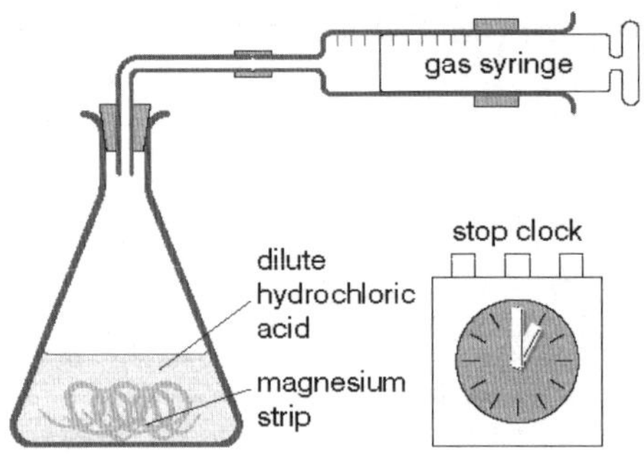

   ii) Length of magnesium strip; concentration and volume of hydrochloric acid.

   iii) In experiment 1 the curve is steeper (has a greater gradient) than in experiment 2.

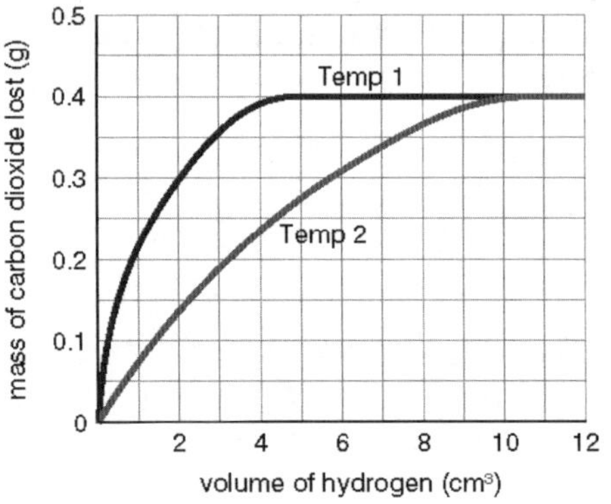

   Experiment 1 – is at the higher temperature. Hydrogen is evolved at a faster rate. Reaction complete as shown (indicated as the volume of gas stays constant).

   Experiment 2 – gradient of graph less showing hydrogen gas evolved at a slower rate reaction complete as shown.

   c) Initially the mass would be the total mass of the container + the unreacted reactants. The rate of loss of mass would be greater at the start of the reaction due to higher concentrations of reactants. The mass would then decrease until the reaction was complete and there would be no more change of mass. If the reactions were monitored at two different temperatures the graphs would start at the same point and lose the same amount of mass. The reaction at the higher temperature would lose the mass at a greater rate.

2. The initial rate of each of the chemical reactions can be found by drawing a gradient to the curve at the starting point (time = 0 seconds).

   A gradient needs to be drawn at the start of the reaction – gradient found as change of Y-axis/change in X-axis:

The initial rate of the reactions for the experiments performed at the same temperatures should be the same as long as the same length of magnesium/concentration and volume of hydrochloric acid are used. The initial rate of the experiment at the higher temperature will be greater and the same for both methods of measurement as long as the conditions were the same. The value for the initial rate of a reaction is independent of the method of monitoring. Units need to be stated.

Activity C4d.3 Consolidation and summary

Learning objectives

- To review the learning points of the topic
- To test understanding through answering questions

Learning outcomes

- Be familiar with the knowledge and understanding summarised in the End of Topic Checklist
- Be able to apply this knowledge and understanding by answering the End of Topic Questions

Resources

Student Book pages 259–260

Approach

This can be short activity coming as it does at the end of a short topic.

Focus on the End of Topic Checklist and Questions in the Student Book.

Answers

End of Topic Questions mark scheme

Question	Correct answer	Marks
1	A reversible reaction is one in which the reactants form the products but the products can reform the reactants.	1 mark
2	In a dynamic equilibrium the reactants are constantly being converted into products and products are constantly being converted back into reactants. The rates of the forward and backward reactions are the same.	1 mark 1 mark
3 a)	Hydrated copper(II) sulfate are blue crystals. On heating steam is given off and condenses as water on the cooler part of the tube. A white/very pale blue powder remains.	1 mark 1 mark 1 mark
3 b)	On adding water steam is produced/the test tube gets warm. A blue solid is formed.	1 mark 1 mark
4	$CuSO_4.5H_2O(s) \rightleftharpoons CuSO_4(s) + 5H_2O(l)$ (1 mark for formulae; 1 mark for state symbols; 1 mark for reversible reaction sign)	3 marks
5	White solid Colourless fumes form on heating. The smoke reforms a white solid on the cooler parts of the tube.	1 mark 1 mark 1 mark
6a)	The equilibrium moves to the right hand side. The equilibrium moves in the direction that reduces the amount of oxygen.	1 mark 1 mark
6 b)	The equilibrium moves to the right hand side. The equilibrium moves in the direction that produces fewer molecules.	1 mark 1 mark
6 c)	The equilibrium moves to the left hand side. The equilibrium absorbs heat energy, the endothermic reverse reaction.	1 mark 1 mark
7 a)	A catalyst increases the rate of a reaction. It provides a reaction pathway with a lower activation energy.	1 mark 1 mark
7 b)	A catalyst does not affect the equilibrium position of a reaction. It increases the rates of the forward and backward reactions to the same extent.	1 mark 1 mark
	Total:	24 marks

Section 4: Exam-style questions mark scheme

Question	Correct answer	Marks
2 a)	i) Correctly plotted graph	3 marks
	Deduct one mark for each incorrectly plotted point.	
	ii) Point at 44 °C circled	1 mark
	iii) Any one sensible source of error: incorrect measurement of temperature, loss in mass or even volume.	1 mark
2 b)	58 ± 2 s	1 mark
	35 ± 2 s	1 mark
2 c)	i) 0.017 ± 0.01	1 mark
	0.029 ± 0.02	1 mark
	ii) g/s	1 mark
	iii) The rate increases.	1 mark
	iv) When the temperature increases the particles will move faster	1 mark
	and collide more frequently.	1 mark
	There will also be more energy involved in the collision and so more collisions will lead to a reaction.	1 mark
2 d)	Put the acid in a fridge or freezer until the temperature falls to 5 °C.	1 mark
3 a)	Energy taken in = +2468 kJ	1 mark
	Energy given out = –2958 kJ	1 mark
	$\Delta H = -490$	1 mark
	kJ/mol	1 mark
3 b)	ii) Axes labelled Energy (kJ) and (Course of) Reaction	1 mark
	Lines for reactants (SO_2 and O_2) and product (SO_3)	1 mark
	ΔH correctly labelled	1 mark
3 c)	i) Pipette used for potassium hydroxide solution	1 mark
	Burette for dilute sulphuric acid	1 mark
	Methyl orange indicator used	1 mark
	Colour change at end point yellow to pink (Phenolphthalein: colour change at end point pink to colourless)	1 mark
	ii) number moles KOH = 0.00375	1 mark
	number of moles of H_2SO_4 = 0.00188	1 mark
	concentration of H_2SO_4 =0 125	1 mark
	M or moles/dm^3	1 mark

Section 5. Chemistry in Society

Contents

a) Extraction and use of metals

b) Crude oil

c) Synthetic polymers

d) The industrial manufacture of chemicals

e) Exam-style questions

Overview of the section

This section applies much of the learning from earlier topics to large-scale industrial processes. In the topic on the extraction and use of metals, the extraction of aluminium from its ore by electrolysis is contrasted with the extraction of iron. The properties of aluminium and iron are then matched to their uses. The topic on crude oil introduces fractional distillation and cracking before considering the environmental impact of burning a fossil fuel such as crude oil or its fractions.

The topic on synthetic polymers contrasts addition and condensation polymerisation, and considers the problems associated with the disposal or recycling of plastic materials. In the last topic, the main focus is on the industrial manufacture of chemicals and the uses of ammonia, sulfuric acid, sodium hydroxide and chlorine.

The topic on extraction and uses of metals builds on previous work on electrolysis and the reactivity series. The topic on crude oil revises covalent bonding and extends the organic chemistry of alkanes and alkenes. The topic on synthetic polymers uses the chemistry of alkenes to introduce addition polymerisation and links to the preparation of alkenes in the cracking of crude oil fractions. Finally, the topic on the industrial manufacture of chemicals is underpinned by an understanding of the factors affecting rate and equilibria and the need to achieve an economic balance between these factors. It also expands on the principles of electrolysis needed to understand the manufacture of sodium hydroxide and chlorine.

In the suggested teaching sequence, all these topics are included near the end of the course. This is because the topics provide not only an extension and application of ideas covered previously but also an opportunity to revise them. As with other sections, it is important to use the 'orientation' stage to establish the extent of students' prior knowledge.

Starting points

The Student Book provides a 'section opener', a double page spread which sets the scene for the students (see pages 266–267). It is structured in this way:

- an overview providing details of the areas of study
- six questions, the first three relating to aspects of chemistry that should be reasonably familiar to most students, and the final three relating directly to some of the topics to be studied
- a list of section contents.

The questions provide a structure for introducing the section.

1. What is the reactivity series? Can you put some of the metals you have studied in an order of reactivity?

2. What are the key features of the alkane homologous series of hydrocarbons?

3. What are anions and cations? How are they discharged at the electrodes in electrolysis?

4. You will be learning about the catalytic cracking of hydrocarbons. What do you think this process might involve?

5. Many polymers are made from alkenes. What do you think will be the important part of the alkene molecule in this polymerisation process?

6. The manufacture of ammonia involves a reversible reaction. In general, what conditions can affect the yield in a reversible reaction?

The section opener has two main purposes:

- to acknowledge the student's prior learning and to value it
- to provide a benchmark against which future learning can be compared.

The six questions or 'starting points' in the Student Book can be used in a number of different ways to introduce the section:

- You could ask students to consider the questions as an introductory homework task.
- You could put students into groups to share their own ideas and understanding and then report back to the whole class.
- Students could be given access to the Internet, preferably with a tight timescale, to find out the information required.

You could then use a spider chart or other form of wall chart to summarise everybody's ideas.

Recording these initial ideas in this way allows you to retain them for reference as the individual topics are developed. In this way, your students' progress in learning can be readily acknowledged.

C5a Extraction and uses of metals

Introduction

This topic is likely to come near to the end of the course. It builds on students' knowledge gained in a number of earlier topics. Topic 2f *Reactivity series* is a good introduction to the extraction of iron from iron(III) oxide, providing a basis for deciding how a metal can be extracted from its ore. Topic 1f *Ionic compounds* and Topic 1i *Electrolysis* give background information for introducing the extraction of aluminium from molten aluminium oxide by electrolysis.

Studying the uses of aluminium and iron in relation to their properties provides an opportunity to recap the structure of a metal as covered in Topic 1h *Metallic crystals*. It is possible to combine parts of this topic with the work on the reactivity series and electrolysis. However, it provides a very good revision of these ideas. This is probably a good reason for leaving the whole topic intact and studying it near the end of the course.

Links to other topics

Sections	Essential background knowledge	Useful links
1 Principles of chemistry	1f Ionic compounds 1h Metallic crystals 1i Electrolysis	
2 Chemistry of the elements	2f Reactivity series	

Topic overview

C5a.1	**Orientation** This activity revises the reactivity series of metals and the significance of carbon's position in the series. The extraction of metals from metal oxides, either by reduction with carbon or by electrolysis of the molten oxide, is then introduced. This is also an opportunity to revise some of the terms related to electrolysis. Students undertake a simple investigation into the reduction of a series of metal oxides using carbon.
C5a.2	**Aluminium: extraction and uses** This activity focuses on using industrial electrolysis to obtain aluminium from bauxite (aluminium oxide). Emphasis is on the use of cryolite, the problems associated with the oxidation of the carbon anodes and the relatively high cost of the process. The reactions at the electrodes are summarised using half equations. Finally, the uses of aluminium are linked to its properties.
C5a.3	**Iron: extraction and uses** This activity focuses on the manufacture of iron in a blast furnace, showing the different stages of the process together with the overall equation for the reduction of the iron(III) oxide. The limitations of the pig iron produced in the blast furnace are highlighted and the subsequent reason for the conversion to steel is emphasised. The uses of the different steel alloys are considered.
C5a.4	**Consolidation and summary** This activity provides an opportunity to recap on the key ideas. Time may be allocated to studying the End of Topic Checklist and answering the End of Topic Questions in the Student Book.

Activity C5a.1 Orientation

Learning objectives

- To revise the concept of the reactivity series
- To relate the method of extraction of a metal to its reactivity
- To revise the reduction of oxides by heating with carbon
- To revise basic ideas of electrolysis

Learning outcomes

- Be able to explain how the methods of extraction of metals relate to their position in the reactivity series
- Know that electrolysis is the breaking down of a compound by passing an electric current through it
- Know that an electrolyte is a substance which allows the passage of an electric current when molten or dissolved in water
- Understand that reduction is loss of oxygen or gain of electrons

Resources

Student Book pages 268–269

Worksheet C5a.1a Obtaining metals from metal oxides

Files on CD–ROM: C5a.1a_worksheet, C5a.1a_tech_notes

Resources for a class practical (see below)

Approach

Ask students what they can tell you about the reactivity series. See if they can construct a series from what they can remember. Refer to the reactivity series in the Student Book and highlight the fact that carbon is included in the series.

Explain that there are three possible ways of obtaining a pure metal:

- Directly from the ground – for example, gold and silver – metals with the lowest reactivity
- By reduction of the metal oxide with carbon – for example copper and iron – metals with mid-range reactivity
- By reduction of the metal oxide using electrolysis – for example, aluminium – the most reactive metals.

There are a number of concepts to revise:

Reduction – loss of oxygen. Refer back to displacement or competition reactions. Give the example of magnesium and copper(II) oxide. Illustrate the redox nature of this reaction.

Reduction – gain of electrons. Refer back to the work on electrolysis. First revise some of the terms: electrode, anode, cathode, electrolyte and electrolysis. Give the example of the electrolysis of lead(II) bromide and write the half equation:

$$Pb^{2+}(l) + 2e^{-} \rightarrow Pb(l)$$

Explain that the lead ions have been reduced to form lead atoms.

Introduce Worksheet 5a.1a. This is a practical activity to reinforce ideas about the reduction of metal oxides, this time using carbon. Ask why carbon is a suitable reducing agent to be used on an industrial scale as opposed to, for example, magnesium.

Technician's notes

You will need the following resources for a class practical.

Be sure to check the latest safety notes on these resources before proceeding.

copper(II) oxide
lead(II) oxide
magnesium oxide, zinc oxide, iron(III) oxide
lime water, carbon powder
eye protection
test tubes, delivery tube, spatula, test tube holder
Bunsen burner

C5a.1a Obtaining metals from metal oxides

Many metals can be extracted from their ores by heating the ore with carbon. The reaction can be summarised by the word equation:

metal oxide + carbon → metal + carbon dioxide

In this reaction the metal oxide is reduced. This only happens if the metal is less reactive than carbon.

Apparatus

test tubes, delivery tube, spatula, test tube holder, Bunsen burner

carbon powder, copper(II) oxide, magnesium oxide, lead(II) oxide, iron(III) oxide, zinc oxide, a solution of lime water

eye protection

SAFETY INFORMATION

Copper(II) oxide is harmful.

Lead(II) oxide is toxic.

Wear eye protection.

Method

1. Put on your eye protection.
2. Mix 1 g spatula measures of carbon powder with 1 g spatula of copper(II) oxide in a dry test tube and set up the apparatus as shown.
3. Heat the test tube strongly for about 2 minutes. **When you stop heating, make sure the delivery tube is removed immediately from the lime water.**
4. When the test tube has cooled down, carefully examine the solid mixture and the lime water and record your results in the table.
5. Repeat the procedure for all the other metal oxides.

NOTE: Room must be well-ventilated.

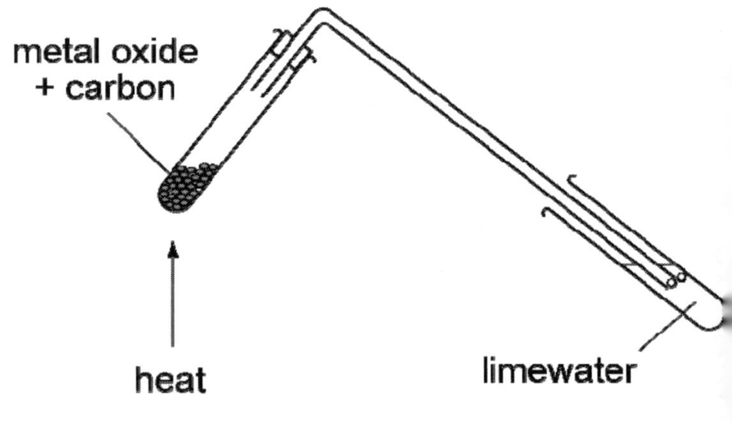

Demonstrate and describe techniques

1. Why is the delivery tube removed from the lime water as soon as the heating is stopped?
2. Why is the limewater used?

Make observations and measurements

Name of metal oxide	Changes in solid mixture	Changes in the lime water
Copper(II) oxide		
Magnesium oxide		
Lead(II) oxide		
Iron(III) oxide		
Zinc oxide		

Analyse and interpret data

1. Which of the metal oxides appear to have been reduced by the carbon? Explain why you have selected each of these metal oxides.

2. Write fully balanced equations, including state symbols, for each of the reactions that lead to the reduction of the metal oxide.

3. Why didn't reactions occur with all five metal oxides?

Activity C5a.2 Aluminium: extraction and uses

Learning objectives

- To study the extraction of aluminium by the electrolysis of aluminium oxide
- To consider the uses of aluminium in relation to its properties

Learning outcomes

- Be able to describe and explain the extraction of aluminium from purified aluminium oxide by electrolysis including:
  the use of molten cryolite as a solvent and to decrease the required operating temperature
  the need to replace the carbon anodes as they react with the oxygen being formed
  the high cost of the electricity
- Know the half-equations for the reactions of aluminium and oxide ions at the cathode and anode electrodes
- Know the uses of aluminium and be able to explain these in terms of its properties

Resources

Student Book pages 269–272

Approach

Introduce the electrolysis of molten aluminium oxide as the means of extracting aluminium from its ore bauxite (aluminium oxide). A diagram of the electrolytic cell used is included in the Student Book. Revisit the terms: electrolysis, electrolyte, electrode, anode and cathode.

Emphasise that the ions in the aluminium oxide must be free to move and that the aluminium oxide must be in a molten state. This is because aluminium oxide is not soluble in water (if aluminium oxide was soluble in water, hydrogen would be discharged in preference to aluminium).

Explain the use of cryolite, the need to replace the carbon anodes and the high cost of electricity. Ask students to write the half equations for the discharge of aluminium and oxide ions.

Consider the uses of aluminium and the properties that make it particularly suited to each use. Encourage students to be precise in describing properties, such as 'low density' rather than 'light' and 'high strength to weight ratio' rather than 'strong'.

Refer to the Science in Context box on pages 271–272 of the Student Book, which includes a section on recycling aluminium.

Answers

Page 271

1. Gold or silver
2. Potassium, sodium, calcium or magnesium
3. Aluminium is above carbon in the reactivity series and so carbon cannot reduce aluminium oxide to aluminium.
4. a) The cryolite produces an electrolyte that has a lower melting point than that of the aluminium oxide and so reduces energy costs.

   b) Aluminium atoms form at the cathode.

   c) $Al^{3+(}l) + 3e^{-} \rightarrow Al(l)$

   d) The aluminium ions are reduced. They accept electrons.

   e) The carbon electrodes react with the oxygen produced in the electrolysis and form carbon dioxide.
5. Aluminium has a high strength to weight ratio/low density/resists corrosion.

Activity C5a.3 Iron: extraction and uses

Learning objectives

- To study the extraction of iron from iron ore in a blast furnace
- To consider the uses of iron in relation to its properties

Learning outcomes

- Be able to describe and explain the main reactions involved in the extraction of iron from iron ore (haematite) using coke, limestone and air in a blast furnace
- Know the uses of iron and be able to explain these in terms of its properties

Common misconceptions

An alloy is a mixture of metals (as opposed to a mixture of a metal with one or more other elements).

Resources

Student Book pages 272–276

Worksheet C5a.3a Extracting iron

Files on CD–ROM: C5a.3a_worksheet

Approach

Introduce the industrial process used to extract iron from its ore, haematite (iron(III) oxide). Ask why electrolysis is not used to extract iron. (Too expensive; reduction with carbon (coke) is much cheaper.)

Consider the diagram of the blast furnace and look at the stages given in the Student Book. Spend some time on the chemical equations for the main reactions. Emphasise that the overall process is reduction.

Reinforce these ideas using Worksheet C5a.3a.

Explain that the iron produced in the blast furnace has limited use – it contains too high a proportion of carbon, which means it is brittle and cannot be easily shaped. Instead, most iron is converted into steel, which is an alloy.

Revise the term 'alloy', a mixture of a metal with another element. By reducing the proportion of carbon in iron obtained from a blast furnace, a much more versatile and useful substance can be made.

Use the Student Book to look at the different compositions of the various steel alloys and their uses.

Answers

Page 275 (top)

1. Iron ore (haematite), coke and limestone
2. Iron(III) oxide
3. Carbon dioxide, carbon monoxide, nitrogen (from the air)
4. $2Fe_2O_3(s) + 3C(s) \rightarrow 4Fe(s) + 3CO_2(g)$
5. a) An alloy is a mixture of a metal and another element.

   b) The proportion of carbon is reduced by heating the pig iron in oxygen.

   c) The pig iron is brittle – steel is more flexible and more resistant to corrosion.

Page 275 (bottom)

1. a) Haematite Fe_2O_3; bauxite Al_2O_3

b) Carbon is not a strong enough reducing agent to remove titanium from its ore. (Note: titanium does form titanium carbide TiC if it is heated with carbon.)

c) i) titanium(IV) oxide + chlorine + carbon(coke) → titanium (IV) chloride + carbon monoxide

$$TiO_2(s) + 2Cl_2(g) + 2C(s) \rightarrow TiCl_4(l) + 2CO(g)$$

ii) titanium(IV) chloride + sodium → titanium + sodium chloride

$$TiCl_4(l) + 4Na(s) \rightarrow Ti(s) + 4NaCl(s)$$

iii) Displacement/redox

iv) Sodium is a reactive metal and will displace titanium from its chloride. Sodium is acting as a reducing agent. Sodium metal will lose electrons to form sodium ions (sodium chloride). Titanium (chloride) will gain electrons to become titanium.

v) Sodium is a very reactive metal (reacts with water and air) the reaction therefore needs to be carried out in an inert atmosphere. Magnesium is still more reactive than titanium and will displace titanium from titanium chloride, but magnesium is not as reactive as sodium so absence of water/air is not as important. (Note, however, $TiCl_4$ does react violently with water)

Reaction $TiCl_4(l) + 2Mg(s) \rightarrow Ti(s) + 2MgCl_2(s)$

vi) Copper is low in the reactivity series. It will not displace titanium from its compounds. Copper is not a strong enough reducing agent to remove titanium from its ore.

C5a.3a Extracting iron

1. Iron is made from iron ore in a blast furnace. There are four raw materials. Draw lines to match them to the main chemical substances in them.

raw material	main chemical substances
haematite (iron ore)	carbon
coke	oxygen and nitrogen
limestone	iron oxide
hot air	calcium carbonate

2. a) Use the words below to label the parts of a blast furnace.

iron ore   molten iron   slag

hot air   coke   limestone

b) Write a word equation to summarise the overall change that happens in a blast furnace.

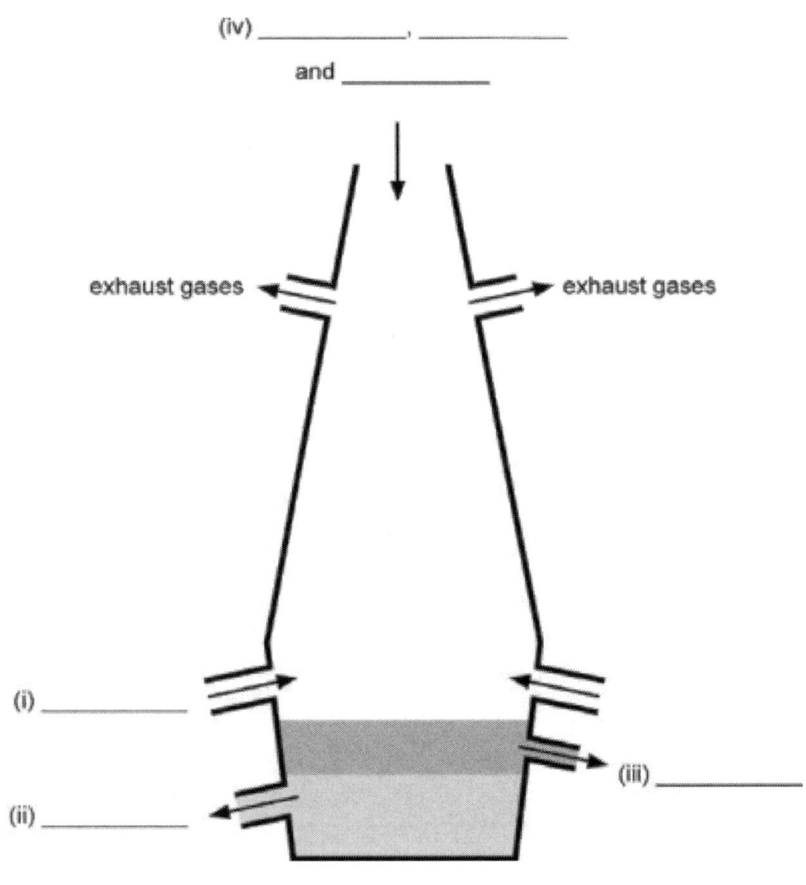

3. a) Match each reaction to its word equations and balanced symbol equation.

Reaction

A carbon monoxide reduces iron oxide

B limestone decomposes to form calcium oxide (which reacts with impurities to form molten slag)

C blast of hot air oxidises coke

Word equation

1. carbon + oxygen → carbon monoxide
2. iron oxide + carbon monoxide → iron + carbon dioxide
3. calcium carbonate → calcium oxide + carbon dioxide

Balanced symbol equation

i) $CaCO_3(s) \rightarrow CaO(s) + CO_2(g)$

ii) $2C(s) + O_2(g) \rightarrow 2CO(g)$

iii) $Fe2O3(s) + 3CO(g) \rightarrow 2Fe(s) + 3CO_2(g)$

b) Explain the difference between an oxidation reaction and a reduction reaction, using reactions in a blast furnace as examples.

Activity C5a.4 Consolidation and summary

Learning objectives

- To review the learning points of the topic
- To test understanding through answering questions

Learning outcomes

- Be familiar with the knowledge and understanding summarised in the End of Topic Checklist
- Be able to apply this knowledge and understanding by answering the End of Topic Questions

Resources

Student Book pages 277–279

Approach

Recap the key learning points of the topic. Focus on the End of Topic Checklist and the End of Topic Questions. Emphasise the aspects that students had forgotten from previous topics of work.

Make sure, in particular, that students are able to write half equations to show the discharge of ions during electrolysis.

Answers

End of Topic Questions mark scheme

Question	Correct answer	Marks
1	Gold and silver are found in their pure state. They are not combined with other elements (as compounds).	1 mark
2 a)	$2Fe_2O_3(s) + 3C(s) \rightarrow 4Fe(s) + 3CO_2(g)$ (1 mark for formulae; 1 mark for balancing)	2 marks
2 b)	The iron(III) oxide is reduced – it has lost oxygen.	1 mark
2 c)	The limestone forms calcium oxide. The calcium oxide combines with impurities such as sand (silicon dioxide) to form a slag (calcium silicate).	1 mark 1 mark
3 a)	$2ZnO(s) + C(s) \rightarrow 2Zn(s) + CO_2(g)$ (1 mark for formulae; 1 mark for balancing) (Also allow: $ZnO(s) + C(s) \rightarrow Zn(s) + CO(g)$)	2 marks
3 b)	The electrolysis requires the zinc oxide to be in a molten state. This requires considerable amounts of energy and so the process is very expensive.	1 mark 1 mark
4 a)	In electrolysis the ions must be free to move (to the electrodes). The ions are able to move in a solution but not in a solid.	1 mark 1 mark
4 b)	$2O^{2-}(l) \rightarrow O_2(g) + 4e^-$ (1 mark for formulae; 1 mark for balancing)	2 marks

Question	Correct answer	Marks
4 c)	The carbon anodes react with the oxygen produced in the reaction. Forming carbon dioxide OR correct equation: $C(s) + O_2(g) \rightarrow CO_2(g)$	1 mark 1 mark
4 d)	$Al^{3+}(l) + 3e^- \rightarrow Al(l)$ (1 mark for formulae; 1 mark for balancing)	2 marks
4 e)	The process has a high demand for electricity (for melting and electrolysis). Hydroelectric power is a relatively cheap source of electricity.	1 mark 1 mark
5 a)	Electrolysis is the breaking down of a compound by an electric current.	1 mark
5 b)	An electrolyte is a substance which when molten or dissolved in water allows the flow of an electric current (and is broken down by it).	1 mark
6 a)	It is an arrangement of ions or atoms (in the case of sodium chloride positive and negative ion) The arrangement extends in 3 dimensions.	1 mark 1 mark
6 b)	The sodium and chloride ions must be free to move (they cannot do this in the solid state).	1 mark
6 c)	Sodium ions will be discharged at the cathode.	1 mark
6 d)	$Na^+(l) + e^- \rightarrow Na(l)$	1 mark
6 e)	Chloride ions will be discharged at the anode.	1 mark
6 f)	$2Cl^-(l) \rightarrow Cl_2(g) + 2e^-$ (1 mark for formulae; 1 mark for balancing)	2 marks
7 a)	An alloy is a mixture of a metal with one or more other elements.	1 mark
7 b)	Steel contains iron (96% approx.) and carbon (4% approx.).	1 mark
8 a)	Like copper, aluminium is a good conductor of electricity. Aluminium is less dense than copper so the cables are lighter.	1 mark 1 mark
8 b)	Aluminium is non-toxic/ impermeable/ resistant to corrosion. Any two.	2 marks
8 c)	High strength to weight ratio/ low density/ resistant to corrosion. Any two.	2 marks
8 d)	Steel is stronger/ less brittle/ more resistant to corrosion than iron. Any two.	2 marks
	Total:	40 marks

C5b Crude oil

Introduction

This topic builds on a number of other topics and is likely to come near the end of the course. Topic 3a *Alkanes* and Topic 3b *Alkenes* previously introduced the idea of saturated and unsaturated hydrocarbons and their combustion in air. Fractional distillation is covered in Topic 1b *Atoms*. In Topic 1g *Covalent substances* students explored the nature of bonding in hydrocarbon molecules and learned the physical properties associated with simple molecular compounds.

An understanding of the formation of oxides, covered in Topic 2d *Oxygen and oxides*, underpins the problems caused by burning substances containing sulfur. The preparation of an unsaturated hydrocarbon by the catalytic cracking of liquid paraffin draws on knowledge of catalysts from Topic 4c *Rates of reaction* and provides a link to Topic 5c *Synthetic polymers*, which covers the preparation of addition polymers from alkenes.

There is some opportunity for investigative skills in this topic.

Links to other topics

Sections	Essential background knowledge	Useful links
1 Principles of chemistry	1b Atoms 1g Covalent substances	
2 Chemistry of the elements	2d Oxygen and oxides	
3 Organic chemistry	3a Alkanes 3b Alkenes	
4 Physical chemistry	4c Rates of reaction	
6 Chemistry in society		5c Synthetic polymers

Topic overview

C5b.1	**Orientation** This activity provides an opportunity to recap ideas on the non-renewable nature of fossil fuels. In addition, you may revise work done previously on the alkane and alkene homologous series of hydrocarbons (such as their saturated/unsaturated characteristics and structural formulae) and so provide a context for studying the fractional distillation of crude oil and catalytic cracking.
C5b.2	**The fractional distillation of crude oil** This activity focuses on the industrial use of the fractional distillation of crude oil. The process is demonstrated and the fractions obtained are compared for colour, viscosity and flammability. Emphasise the nature and uses of the major fractions obtained from the fractional distillation.
C5b.3	**Cracking crude oil fractions** The reason for the importance of the catalytic cracking of the 'heavier' crude oil fractions is emphasised (converting the less useful longer-chain hydrocarbons into more useful shorter-chain hydrocarbons). Students have the opportunity to crack liquid paraffin and to demonstrate that an alkene is formed in this process. Practice is provided in writing structural formulae for reactants and products (alkanes and alkenes).

C5b.4	**The combustion of fuels and acid rain** This activity revises the combustion reactions of hydrocarbons, contrasting the products formed if there is a plentiful supply of air with those if the air supply is limited. The focus then turns to acid rain, how it forms in the atmosphere and the damage it can cause to the environment. Ways of reducing the formation and impact of acid rain are then considered.
C5b.5	**Consolidation and summary** This activity considers the advantages and disadvantages of using fossil fuels before recapping the key ideas encountered in the topic. Time can be allocated to studying the End of Topic Checklist and then answering the End of Topic Questions in the Student Book.

Activity C5b.1 Orientation

Learning objectives

- To introduce crude oil as a fossil fuel
- To revise ideas on the alkane and alkene series of hydrocarbons

Learning outcomes

- Understand how crude oil is formed and that it is a non-renewable fossil fuel
- Know that a hydrocarbon is a compound containing hydrogen and carbon only
- Know that alkanes are a series of saturated hydrocarbons
- Know that alkenes are a series of unsaturated hydrocarbons

Resources

Student Book pages 280–281

Approach

Ask students to name some of the common fossil fuels and to explain why they are non-renewable. Explain that crude oil is a mixture of a large number of organic compounds, many of which are used as fuels.

Remind students about the alkane homologous series of saturated hydrocarbons. Ask them to explain what a saturated hydrocarbon is. Ask what an unsaturated hydrocarbon is, and ask for some examples from the alkene homologous series.

Ask students to draw displayed formulae for butane and butene to illustrate the differences in structure. This part of the activity should help you to gauge students' knowledge of alkanes and alkenes and may offer an opportunity for a more detailed revision.

Activity C5b.2 The fractional distillation of crude oil

Learning objectives

- To study the industrial process of the fractional distillation of crude oil
- To consider the nature and properties of the different substances (fractions) that can be obtained from crude oil

Learning outcomes

- Understand that crude oil is a mixture of hydrocarbons
- Be able to describe and explain how the industrial process of fractional distillation separates crude oil into fractions
- Know the names and uses of the main fractions obtained from crude oil:
  refinery gases
  gasoline
  kerosene
  diesel
  fuel oil
  bitumen
- Be able to describe the trend in boiling point and viscosity of the main fractions

Resources

Student Book pages 281–283

Worksheet C5b.2a The fractional distillation of crude oil

Files on CD–ROM: C5b.2a_worksheet, C5b.2a_tech_notes

Resources for a demonstration (see below)

Approach

Ask students what process can be used to separate a mixture of liquids (fractional distillation). Explain that crude oil can be separated into components using fractional distillation, but rather than separating liquid by liquid (or compound by compound), it is separated into fractions with a range of boiling points which are therefore likely to contain more than one compound.

Demonstrate the fractional distillation of a crude oil substitute. Explain that crude oil is not allowed in schools but that the technician has made up a substitute, which is much less dangerous than crude oil itself, though still harmful. Use Worksheet C5b.2a so that students can record the observations and measurements and analyse their findings. Trends in boiling points, colour, viscosity, ease of burning and smokiness of the flame for the fractions are given in the Student Book pages 282–283.

It is important that students know some of the key components of the fractions (refinery gases, gasoline, kerosene, diesel, fuel oil and bitumen), where they are produced in the fractionating column and the approximate number of carbon atoms they have in their molecules.

Technician's notes

You will need the following resources for a demonstration.

Be sure to check the latest safety notes on these resources before proceeding.

100 cm^3 crude oil substitute made from: • 55 cm^3 liquid paraffin, 20 cm^3 paraffin oil (kerosene) • 11 cm^3 white spirit; 4 cm^3 petroleum ether (100–120 °C) • 4 cm^3 of petroleum ether (80–100 °C); 6 cm^3 of petroleum ether (60–80 °C) • black oil paint (to provide colour)
Bunsen burner and heat resistant mat, mineral wool, tongs
side arm Borosilicate glass test tube, delivery tube and rubber connection tubing
thermometer (0–360 °C), teat pipette, beaker, watch glass
eye protection

Ideally the demonstration should be undertaken in a fume cupboard.

Crude oil must not be used in schools.

Answers

Page 283

1. The supplies of crude oil are limited – it takes millions of years for crude oil to be formed.
2. Natural gas or methane. It is trapped in pockets above the oil.
3. Small chain of carbon atoms.
4. Long chain of carbon atoms.
5. These fractions ready form a vapour.

C5b.2a The fractional distillation of crude oil

Crude oil is a mixture of many different liquids. Because these liquids have different boiling points, they can be separated by fractional distillation.

You will be shown a demonstration of how crude oil can be separated into fractions with different boiling points. These fractions will then be compared to see how their properties vary.

Apparatus

Bunsen burner and heat resistant mat, mineral wool, tongs

side arm Borosilicate glass test tube, delivery tube and rubber connection tubing

thermometer (0–360 °C), teat pipette, beaker, watch glass

crude oil substitute

eye protection

SAFETY INFORMATION

Crude oil substitute is highly flammable and harmful.

Method

1. Put on your eye protection.
2. The crude oil will be heated slowly and four or five different fractions collected with boiling points in different temperature ranges.
3. The different fractions will then be observed and tested. Record your observations in the table below.

Note: viscosity is a measure of the runniness (mobility) of the liquid.

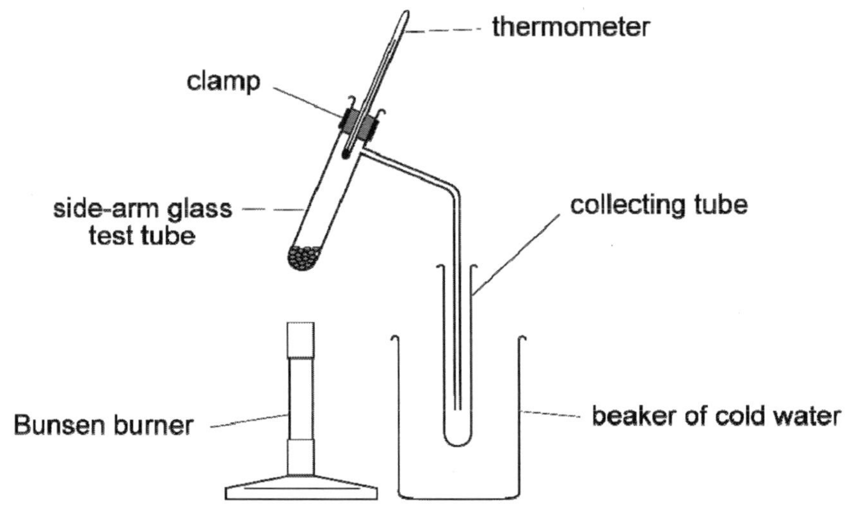

Page 1 of 2

Make observations and measurements

Fraction	Boiling point range (°C)	Colour	Viscosity	Observations on burning (flammability)
1				
2				
3				
4				
5				

Analyse and interpret data

1. Summarise the trends in properties from fraction 1 to fraction 5.

Property	Trend from fraction 1 to fraction 5
Colour	
Viscosity	
Flammability	

2. What colour flame indicates that a hydrocarbon fuel has sufficient oxygen to burn completely?
3. What products are formed when a hydrocarbon fuel burns completely in oxygen?
4. What colour flame indicates that a hydrocarbon fuel has insufficient oxygen to burn completely?
5. What products could be formed when a hydrocarbon fuel burns in an insufficient supply of oxygen?

Activity C5b.3 Cracking crude oil fractions

Learning objectives

- To introduce the process of cracking hydrocarbon chains
- To explain the reasons for using the cracking process

Learning outcomes

- Understand that the fractional distillation of crude oil produces more long-chain hydrocarbons than can be used and fewer short-chain hydrocarbons than are needed and so cracking is necessary
- Be able to describe how long-chain hydrocarbons can be converted into shorter chain hydrocarbons by catalytic cracking
- Know that oil fractions are cracked using catalysts of silica or alumina at a temperature of between 600 and 700 °C

Resources

Student Book pages 283–285

Worksheet C5b.3a Cracking liquid paraffin

Files on CD–ROM: C5b.3a_worksheet, C5b.3a_tech_notes

Resources for a class practical (see below)

Approach

Refer to Table 5.3 on page 283 of the Student Book which shows the typical percentages of the different fractions produced by the fractional distillation of crude oil. Explain that the fractions most easily used are the 'lighter' fractions, such as gasoline (13%), and so a process is needed to convert the 'heavier' fractions into the lighter fractions.

Emphasise that one of the products of cracking is always an alkene and that alkenes are much more reactive than alkanes, so they can be converted (by addition reactions) into a much wider range of useful chemicals.

Worksheet C5b.3a provides details of a class practical involving the cracking of liquid paraffin. The students will need to be warned to be careful that a 'suck back' doesn't take place. This process also features in the *Developing Investigative Skills* in the Student Book on pages 284–285.

You will need to decide whether to refer to both and, if you decide to, in what order you will introduce them. Students may need to be reminded of the use of bromine water to identify an unsaturated hydrocarbon (alkene).

Take time to practise writing structural formulae for the alkanes and alkene involved in the reaction of cracking.

Technician's notes

The following resources are needed for a class practical.

Be sure to check the latest safety notes on these resources before proceeding.

medicinal paraffin
bromine water (0.02M)
pot fragments
Bunsen burner and heat resistant mat
stand and clamp, boiling tube, bung and delivery tube
mineral wool and tongs
water basin, test tubes, 2 dropping pipettes
eye protection

Answers

Developing investigative skills, page 284

1. The gases must be insoluble in water.

2. To prevent a 'suck back' (where water is forced back through the apparatus as the gases cool and contract) and water reaching the boiling tube that has been heated.

3. a) The liquid paraffin is flammable. Care must be taken to make sure the apparatus is assembled securely.

   b) The boiling tube could crack while it is being heated. Safety glasses should be worn.

   c) There is danger of a 'suck back' when heating is stopped. This can be prevented by removing the delivery tube from the trough of water.

4. The first test tube would contain a large volume of air displaced from the apparatus.

5. The gas (hydrocarbon) must be unsaturated, that is, it contains a carbon-to-carbon double bond.

6. $C_{14}H_{30}(g) \rightarrow C_{12}H_{26}(g) + C_2H_4(g)$

Page 285

1. Ethene is a member of the alkene homologous series.

2. The fractional distillation of crude oil produces a high proportion of long-chain hydrocarbons, which are not as useful as short-chain hydrocarbons. Cracking converts the long-chain hydrocarbons into more useful shorter chain hydrocarbons.

3. The conditions required for cracking oil fractions are a temperature of between 600 and 700 °C and a catalyst of silica or alumina.

C5b.3a Cracking liquid paraffin

Liquid paraffin contains long-chain alkanes. These long-chain alkanes can be broken down into shorter chain alkanes and alkenes by a process known as cracking. Cracking needs a high temperature and a catalyst to give a suitable rate of reaction. In this experiment pieces of pot will act as the catalyst.

Apparatus

pot fragments

Bunsen burner and heat resistant mat

stand and clamp, boiling tube, bung and delivery tube

mineral wool, tongs, two dropping pipettes, water basin, test tubes

liquid paraffin, bromine water

eye protection

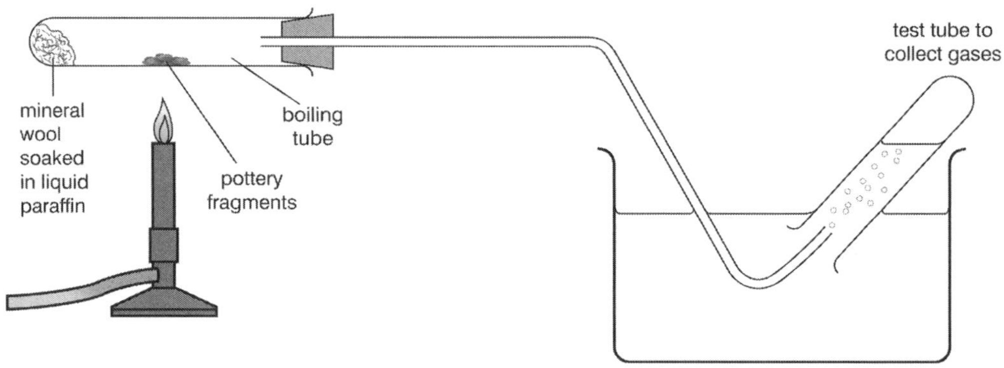

SAFETY INFORMATION
Bromine water is harmful.
Wear eye protection.

Method

1. Put on your eye protection.
2. Put some mineral wool in the bottom of a boiling tube. Using a dropping pipette, a few drops of liquid paraffin to soak the mineral wool.
3. Clamp the boiling tube horizontally as shown in the diagram and place a pile of pot fragments in the centre of the tube. Then complete the setting up of the apparatus.
4. Fill three or four test tubes with water and place them in the trough ready to collect the gas.
5. Heat the pot fragments strongly and occasionally heat the mineral wool for a second or two.
6. Collect the gas produced in the test tubes.
7. When all the test tubes are full of gas, **remove the delivery tube from the water and then stop heating.**

Devise and plan investigations

1. Why is the delivery tube removed from the water before heating is stopped?

Make observations and measurements

1. Test the gas you have produced and record your observations in the table.

Test	Observations
Add a lighted splint	
Add 5 or 6 drops of bromine water	

Analyse and interpret data

1. What conclusions can you make about the gas you collected in the experiment?

Activity C5b.4 The combustion of fuels and acid rain

Learning objectives

- To consider some of the problems of pollution caused by burning hydrocarbon fuels
- To consider specifically the problems of acid rain

Learning outcomes

- Understand that the incomplete combustion of hydrocarbon fuels may produce carbon monoxide
- Be able to explain that carbon monoxide is poisonous because it reduces the capacity of the blood to carry oxygen
- Understand that in car engines the temperature is hot enough for nitrogen and oxygen to combine and form nitrogen oxides
- Understand that nitrogen oxides and sulfur dioxide are pollutant gases which contribute to acid rain, and be able to describe the problems caused by acid rain

Resources

Student Book pages 285–290

Approach

Students should be familiar with the combustion of hydrocarbons and the formation of carbon dioxide and water when the supply of oxygen is plentiful. The sign that a fuel has enough air supply to burn completely is that the flame will be blue. Students may also remember that the formation of carbon or carbon monoxide indicates that supplies of oxygen are limited – indicated by a yellow flame.

Students will be familiar with the production of carbon dioxide from the burning of carbon and of sulfur dioxide from burning sulfur in air or oxygen. Explain that sulfur is present in fossil fuels such as coal, oil and gas and so sulfur dioxide is formed when these fuels are burned.

Use Student Book page 287 to consider the equations for converting sulfur dioxide into sulfuric acid. Similarly, consider the formation of nitric acid from oxides of nitrogen, which are also produced when burning fuels and particularly in car engines.

Explain that the formation of sulfur dioxide and the oxides of nitrogen is largely responsible for the formation of acid rain. Refer to pages 288–289 in the Student Book as a way of discussing the impact of acid rain and the ways it can be reduced. Ask students to summarise the different impacts of acid rain, the ways that its production can be reduced and strategies for reducing damage by acid rain.

Answers

Page 287

1. The fuel will burn with a yellow (rather than blue) flame.

2. Carbon and carbon monoxide.

3. It combines with the haemoglobin to form carboxyhaemoglobin, which prevents the haemoglobin from combining with oxygen.

4. Wind, wave, solar and nuclear power are alternative ways of generating energy.

Page 289

1. The sulfur dioxide combines with oxygen and water to form sulfuric acid.

2. Nitrogen oxide reacts with oxygen and water to form nitric acid.

3. An alkali such as slaked lime.

Activity C5b.5 Consolidation and summary

Learning objectives

- To review the learning points of the topic
- To test understanding through answering questions
- To consider sources of energy for the future

Learning outcomes

- Be able to explain the limitations of using fossil fuels as sources of energy
- Be familiar with the knowledge and understanding summarised in the End of Topic Checklist
- Be able to apply this knowledge and understanding by answering End of Topic Questions

Common misconceptions

Some students think that all fossil fuels will run out soon. (Supplies of coal are relatively large.)

Resources

Student Book pages 289–294

Approach

Use the Science in Context box *The Fossil Fuel Dilemma* on pages 289–290 of the Student Book to consider the advantages and disadvantages of using fossil fuels to generate energy. Are alternative forms of energy the answer? Should coal be used more extensively?

Recap the key learning points of the topic. Focus on the End of Topic Checklist and End of Topic Questions. Check the students have a good understanding of the aspects of organic chemistry encountered in the topic.

Answers

End of Topic Questions mark scheme

Question	Correct answer	Marks
1 a)	Crude oil is formed from the remains of animals. Pressed together under layers of rock (over millions of years).	1 mark 1 mark
1 b)	It takes millions of years to form – crude oil used cannot be replaced.	1 mark
2 a)	Fractional distillation	1 mark
2 b)	The boiling point decreases	1 mark
2 c)	Some parts of the mixture condense and form liquids (which are drained off). The parts that do not condense (temperature above the boiling point) pass up the column.	1 mark 1 mark
3 a)	A hydrocarbon is a substance that contains carbon and hydrogen atoms only.	1 mark
3 b)	A catalyst of silica or alumina A temperature of between 600 and 700 °C	1 mark 1 mark
3 c)	C_8H_{18}	1 mark
3 d)	The alkane series	1 mark
3 e)	Cracking converts the less useful longer chain molecules into more useful shorter chain molecules.	1 mark 1 mark
4 a)	Carbon dioxide Water	1 mark 1 mark
4 b)	$2C_8H_{18}(l) + 25O_2(g) \rightarrow 16CO_2(g) + 18H_2O(l)$ (1 mark for formulae; 1 mark for balancing)	2 marks
4 c)	Carbon monoxide is highly poisonous. It reduces the capacity of the blood to transport oxygen around the body.	1 mark 1 mark
5 a)	Sulfur dioxide/nitrogen oxide/carbon dioxide. Any two.	2 marks
5 b)	It damages plants/animals that feed on the plants/leaches minerals out of the soil. Any two.	2 marks
5 c)	Burn less fossil fuels/use filters to remove gases such as SO_2 at power stations/add lime to lakes and rivers. Any two.	2 marks
	Total:	26 marks

C5c Synthetic polymers

Introduction

This is a short topic that most appropriately follows on from a study of Topic 1g *Covalent substances*, Topic 3a *Alkanes* and Topic 3b *Alkenes*. It provides an opportunity to revise ideas on covalent bonding in organic molecules, as well as specific revision of the addition reactions of alkenes.

After this revision, the topic focuses on addition polymerisation and then condensation polymerisation. The differences between the two processes are emphasised. There are opportunities to consider some of the environmental aspects of the manufacture and use of synthetic polymers, particularly recycling and disposal. There are some limited opportunities for students to develop their investigative skills.

Links to other topics

Sections	**Essential background knowledge**	**Useful links**
1 Principles of chemistry	1g Covalent substances	
3 Organic chemistry	3a Alkanes 3b Alkenes	

Topic overview

C5c.1	**Orientation** This activity provides an opportunity to revise the organic chemistry of alkanes and alkenes. The terms 'monomer', 'polymer' and 'polymerisation' are introduced.
C5c.2	**Addition polymerisation** This activity focuses on how single alkene monomers, or substituted alkene monomers, can combine to form addition polymer molecules. It emphasises how the monomers combine and the resulting repeating topic of polymers. The reverse process identifies the monomer molecule from the structural formula of the polymer. Examples are given of common addition polymers and their uses, and the problems posed by the non-biodegradability of many addition polymers.
C5c.3	**Condensation polymerisation** This is a short activity introducing condensation polymerisation as a contrast to addition polymerisation. Little detail is provided of the structures of the monomers, instead the emphasis is placed on the formation of a small molecule (often water) when two different monomer molecules combine.
C5c.4	**Consolidation and summary** This activity is an opportunity to recap on the key ideas of the topic. Time can be allocated to studying the End of Topic Checklist and then answering the End of Topic Questions in the Student Book.

Activity C5c.1 Orientation

Learning objectives

- To revise the structures of the alkanes and alkenes
- To revise the addition reactions of the alkenes
- To introduce the terms monomer, polymer and polymerisation

Learning outcomes

- Understand that alkenes are much more reactive than alkanes
- Understand that alkenes undergo addition reactions
- Know that a monomer is a molecule that can combine with other molecules to form a polymer

Resources

Student Book pages 295–296

Approach

Ask students to draw the structural formulae of the first three members of the alkane series. Check that they can remember the names of the alkanes and that they know alkanes are saturated hydrocarbons.

Repeat the questioning with the first three members of the alkene series. Check that students know that alkenes are unsaturated hydrocarbons.

Ask what reaction is used to identify an alkene or unsaturated hydrocarbon (adding bromine water). Check that students remember that this is an addition reaction.

Introduce the terms 'monomer', 'polymer' and 'polymerisation'. Explain that polymers are often referred to as plastics. Ask students, in groups, to make a list of as many things as they know about polymers – allow only a few minutes to do this.

Review the lists. Disposal of plastics and recycling will be common emerging themes.

Explain that the topic focuses on two different ways of making synthetic polymers. The first method involves alkenes and makes use of their reactivity and ability to undergo addition reactions. The second method involves a very different type of reaction. The disposal and recycling of plastics will also be considered.

Activity C5c.2 Addition polymerisation

Learning objectives

- To introduce and explain the nature of addition polymerisation.
- To consider the uses of some of the common examples of addition polymers.
- To consider the problems of recycling and disposing of used plastic material.

Learning outcomes

- Understand that an addition polymer is formed by joining up many small molecules called monomers.
- Be able to draw repeat topics of addition polymers from the structural formula of the monomer, such as poly(ethene), poly(propene) and poly(chloroethene).
- Be able to deduce the structural formula of a monomer from the repeat topic of an addition polymer.
- Be able to describe some of the uses for polymers, including poly(ethene), poly(propene) and poly(chloroethene).
- Be able to explain that addition polymers are hard to dispose of as their inertness means that they do not easily biodegrade.

Resources

Student Book pages 295–297

Worksheet C5c.2a Identifying addition polymers

Files on CD–ROM: C5c.2a_worksheet, C5c.2a_tech_notes

Resources for a class practical (see below)

Approach

Introduce the definition of an addition polymer and use structural formulae to illustrate how alkene molecules can undergo addition reactions to form a polymer chain. Show how substituted alkenes such as chloroethene can similarly form addition polymers.

Use page 296 of the Student Book to show some of the addition polymers in common use. Give students practice at deducing the repeating topic of an addition polymer from the structure of the monomer and then the reverse – that is, working out the structure of the monomer from the structural formula of part of an addition polymer. Consider the uses of the common addition polymers.

Use Worksheet C5c.2a to support a class practical activity in which students identify three different samples of addition polymers or plastics, essentially by working out the approximate density of the plastic sample.

Rehearse some of the familiar problems of using plastics, both in terms of their disposal and recycling. Emphasise and explain the non-biodegradability of many addition polymers.

Refer to the three Rs (Reduce, Reuse, Recycle) and ask the students how these three Rs are being used, or could be used more effectively, to prevent waste. Use the Science in Context section *The challenges of recycling plastics* on page 298 of the Student Book to explain that recycling is complicated, not just because different plastic materials need to be separated but also because some plastic materials cannot be melted down and recycled (thermosetting plastics).

Technician's notes

You will need the following resources for a class practical.

Be sure to check the latest safety notes on these resources before proceeding.

liquid 1 (ethanol)
liquid 2 (596 cm^3 ethanol + 439 cm^3 distilled water)
liquid 3 (448 cm^3 ethanol + 586 cm^3 distilled water)
liquid 4 (distilled water)
liquid 5 (184 g of potassium carbonate dissolved in 965 cm^3 distilled water)
liquid 6 (513 g potassium carbonate dissolved in 866 cm^3 distilled water)
scissors, boiling tubes, boiling tube racks, glass stirring rod
plastic samples labelled A, B and C
eye protection

Answers

Page 297

1. The individual beads are like monomer molecules. The string of beads is like a polymer made by joining together many of these monomers.

2. Poly(ethene)/polythene is used to make plastic bags.

3. a)

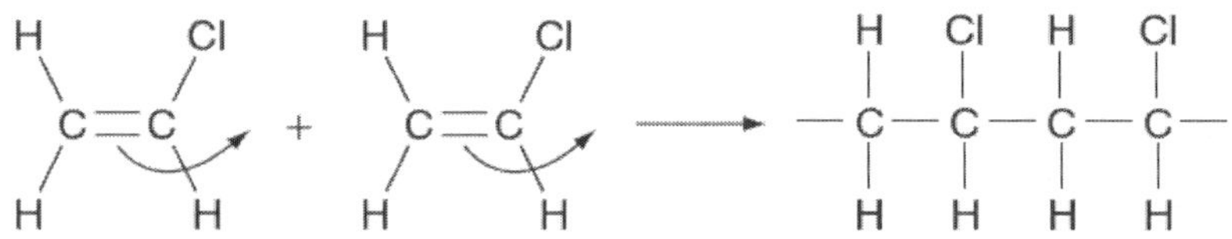

b)

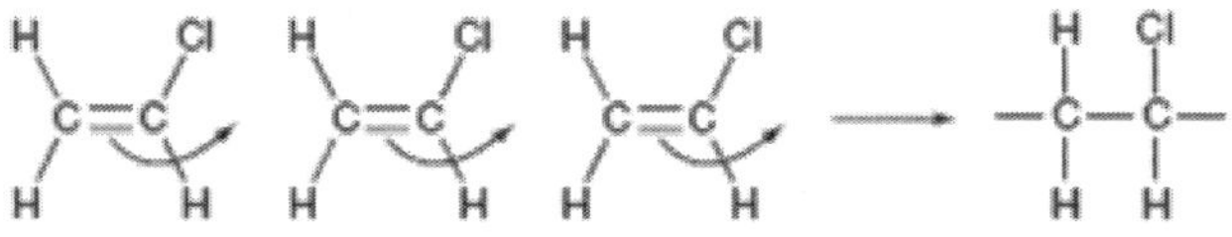

4. Poly(chloroethene) is an addition polymer.

C5c.2a Identifying addition polymers

There is a large number of addition polymers in everyday use, as the table below shows:

Polymer	Alternative name	Density (g/cm^3)	Common use
Expanded poly(phenylethene)	Expanded polystyrene	0.02–0.06	Burger box
Poly(propene)	Polypropylene	0.89–0.91	Low fat soft cheese tub
Low density poly(ethene)	Low density polythene (LDPE)	0.91–0.93	Branded lemon juice container
High density poly(ethene)	High density polythene (HDPE)	0.94–0.96	Fabric conditioner bottle
Poly(phenylethene)	Polystyrene	1.04–1.11	Yoghurt pot
Poly(chloroethene)	Polyvinylchloride (PVC)	1.20–1.55	Shampoo bottle

You are going to use a 'float' test to help identify 3 unknown samples of addition polymers or plastics.

Apparatus

scissors, boiling tubes, boiling tube rack, glass stirring rod

samples of plastic A, B and C

SAFETY INFORMATION

Liquids 1, 2 and 3 are highly flammable.

Liquids 5 and 6 are irritants.

Wear eye protection.

Method

1. Put on your eye protection.
2. Pour 20 cm^3 of each of the liquids 1-6 into a separate boiling tube.
3. Cut out samples of the plastics A, B and C into 4 mm × 4 mm squares. Add a sample of A to each of the liquids. Push the sample with the stirring rod to see whether it sinks or floats.
4. Repeat the procedure with samples B and C. Record your results in the table below.

Page 1 of 2

Make observations and measurements

Liquid	Density (g/cm^3)	Observations on adding plastic (√ = floats; x = sinks)		
		A	B	C
1	0.79			
2	0.91			
3	0.94			
4	1.00			
5	1.15			
6	1.38			

Analyse and interpret data

Use your observations and the table of information at the beginning of the worksheet to identify the plastics in samples A, B and C.

Plastic sample	Name of the addition polymer in the plastic material
A	
B	
C	

Activity C5c.3 Condensation polymerisation

Learning objectives

- To study the formation of condensation polymers
- To consider the differences between condensation and addition polymers

Learning outcomes

- Understand that some polymers, such as nylon, form by a different process called condensation polymerisation
- Understand that when condensation polymers are formed small molecules, such as water, are released

Resources

Student Book pages 297–299

Files on CD–ROM: C5c.3a_tech_notes

Resources for a demonstration (see below)

Approach

You may take a simple approach to condensation polymerisation, since the structural formulae of the monomer molecules are complicated, typically involving functional groups beyond the scope of this course. Simple 'block' representations can be used for the two different monomers involved in the condensation process.

Demonstrate the nylon 'rope trick' to show how nylon can be made from two monomers – referred to simply as monomer A and monomer B. Explain that the term 'condensation' results from the fact that water is often formed when the two monomers combine together. Contrast this with addition polymerisation, where the only product is the polymer.

Technician's notes

You will need the following resources for a demonstration.

Be sure to check the latest safety notes on these resources before proceeding.

solution A (2.2 g 1,6-diaminohexane in 50 cm^3 distilled water)
solution B (1.5 g decanedioyl chloride (sebacoyl chloride) in 50 cm^3 of cyclohexane
50 cm^3 beaker, tweezers
eye protection

Answers

Page 298

1. Nylon is made from two monomers and a small molecule is eliminated when the two monomers combine. An addition polymer has only one monomer.

2. To produce a polymer chain the monomers need to be able to join together at both ends of the molecules.

Page 299

1. Fractional distillation: can be used to separate crude oil into different fractions having different boiling points. (Note more detail may be included here – see 'How does fractional distillation work?')
Cracking: is the splitting up of a large chain hydrocarbon into shorter more useful one.
Polymerisation: There are two types of addition polymerisation and condensation polymerisation. Addition polymerisation is the formation of a polymer formed by the addition which occurs across carbon-carbon double bonds when monomer units combine, such as ethene to poly(ethene). Condensation polymerisation is the formation of a polymer in which the combination of two substances is accompanied by the elimination of a small simple molecule such as water or hydrogen chloride: for example, nylon is a condensation polymer.

2. a) Heptane C_7H_{16}

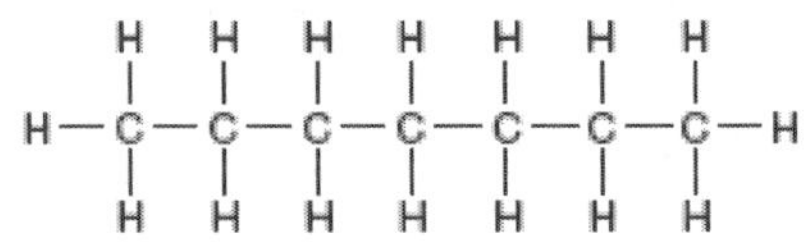

b) Heptane forms part of a fraction called gasoline which is a very volatile low boiling point liquid. The state of alkane will be determined by the length of the carbon–carbon chain and the strength of the intermolecular forces. The larger the molecules, the stronger the intermolecular forces. The simplest alkanes are gases, the longer chains are liquids and even longer 20+ C solids heptane is a liquid.

c) $C_7H_{16} \rightarrow C_5H_{12} + C_2H_4$ (any combination of alkanes and alkenes which add up to 7C and 16H)

d) Step 1 formation of ethene. By the cracking of any long chain alkane to give a product of ethene (C_2H_4) (any balanced equation of alkane $\rightarrow$ shorter alkane + C_2H_4)

Step 2 formation of chloroethane: By an addition reaction between ethene and HCl

$CH_2{=}CH_2 + HCl \rightarrow CH_3CH_2Cl$

Step 3 formation of PVC: By the addition polymerisation of chloroethene to form poly(chloroethene)

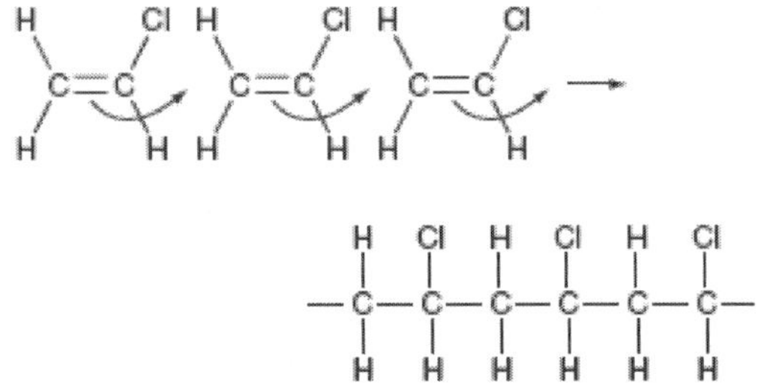

Activity C5c.4 Consolidation and summary

Learning objectives

- To review the learning points of the topic
- To test understanding through answering questions

Learning outcomes

- Be familiar with the knowledge and understanding summarised in the End of Topic Checklist
- Be able to apply this knowledge and understanding by answering the End of Topic Questions

Resources

Student Book pages 300–301

Approach

Recap the key learning points of the topic. Focus on the End of Topic Checklist and End of Topic Questions. Take the opportunity to revise alkanes and alkenes, particularly if students' recall of these was limited at the start of this topic.

Answers

End of Topic Questions mark scheme

Question	Correct answer	Marks
1 a)	A monomer is a molecule that can combine with other molecules to form a polymer.	1 mark
1 b)	A polymer is a giant (large) molecule made up of one or more monomers.	1 mark
2 a)	An addition polymer is formed when molecules of a single monomer join together in large numbers.	1 mark
2 b)	A C=C double bond	1 mark
2 c)	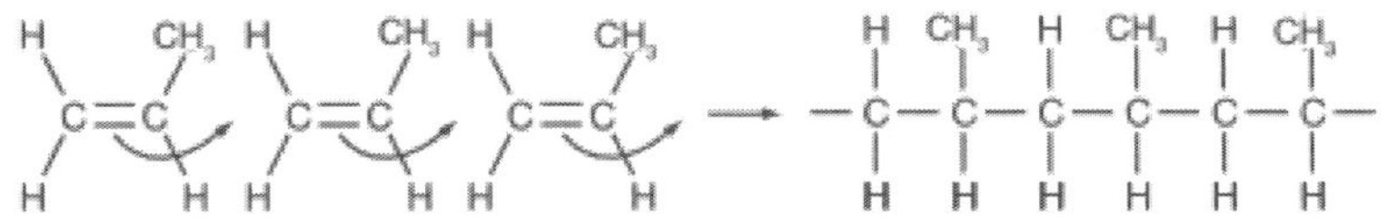	2 marks
2 d)	Propane is a saturated hydrocarbon and does not contain a C=C double bone. It cannot undergo addition reactions and so cannot form a polymer.	1 mark 1 mark
3	Completed table as shown below.	3 marks

Addition polymer	Use	
Poly(ethene)	Buckets/bowls/plastic bags	Any one. (1 mark)
Poly(propene)	Packaging/ropes/carpets	Any one. (1 mark)
Poly(chloroethene)	Plastic sheets/artificial leather	Any one. (1 mark)

4 a)	Addition polymers cannot be broken down. By the bacteria in the soil.	1 mark 1 mark
4 b)	Plastics have to be separated from other recyclable materials. There are different types of plastic – they need to be separated. Only some (thermoplastic) plastics can be melted and reformed others (thermosetting) decompose on strong heating.	1 mark 1 mark 1 mark
5 a)	A condensation polymer is made from two different monomers. A small molecule (such as water) is also produced when two monomers combine.	1 mark 1 mark
5 b)	i) Nylon/terylene/allow polyester/polyurethane. Any one. ii) Nylon/terylene/polyester used in clothing (accept other sensible alternatives).	1 mark 1 mark
	Total:	20 marks

C5d The industrial manufacture of chemicals

Introduction

The topic has three main sections: the manufacture of ammonia, the manufacture of sulfuric acid and the manufacture of sodium hydroxide and chlorine.

There are very direct links to a number of topics. This means that you could teach the topic in three separate parts, with the sections on ammonia and sulfuric acid (equilibrium reactions with a balance being sought between yield and rate) alongside Topic 4c *Rates of reaction* and Topic 4d *Equilibria*; and the section on sodium hydroxide and chlorine (electrolysis in a diaphragm cell) taught alongside Topic 1i *Electrolysis*. The advantage of this approach would be teaching the industrial applications close to the underpinning chemical principles.

Alternatively, you might use this topic as a means of revising and consolidating previous learning. A possible drawback of this approach is that the topic provides few opportunities for investigative practical work. The oxidation of nitrogen monoxide in the presence of water to nitric acid provides a link to the formation of acid rain, covered in Topic 5b *Crude oil*.

Links to other topics

Sections	Essential background knowledge	Useful links
1 Principles of chemistry	1i Electrolysis	
2 Chemistry of the elements	2d Oxygen and oxides	
4 Physical chemistry	4a Acids, alkalis and salts 4b Energetics 4c Rates of reaction 4d Equilibria	
5 Chemistry in society		5b Crude oil

Topic overview

C5d.1	**Orientation** This activity will only be needed if you teach the topic as a whole. It enables students to review and revise earlier underpinning work on redox reactions, rates of reaction, energetics, equilibria and electrolysis, some of which relate to each of the three manufacturing processes being considered.
C5d.2	**The manufacture of ammonia** This activity focuses on the reaction and conditions for manufacturing ammonia. It highlights the need to achieve a balance between obtaining a good yield of ammonia and doing so in a suitable time. Ammonia's importance is emphasised by considering its use in the manufacture of nitric acid and fertilisers.
C5d.3	**The manufacture of sulfuric acid** This activity mirrors C5d.2, but this time with the focus on sulfuric acid. It highlights the need to balance between obtaining a good yield of sulfuric acid and doing so in a suitable time. The importance of sulfuric acid is emphasised by considering its use in the manufacture of detergents, fertilisers and paints.

C5d.4	**The manufacture of sodium hydroxide and chlorine** This activity is based on the electrolysis of concentrated sodium chloride solution in the diaphragm cell. Students should understand electrolysis and have the opportunity to see how this can be applied on an industrial scale. The uses of sodium hydroxide and chlorine are also considered.
C5d.5	**Consolidation and summary** This activity is only needed if the topic is taught as a whole. It provides an opportunity to recap on the ideas encountered in the topic. Time can be allocated to studying the End of Topic Checklist and then answering the End of Topic Questions in the Student Book.

Activity C5d.1 Orientation

Learning objectives

- To revise ideas on reaction rate and equilibria
- To revise understanding of electrolysis
- To revise redox reactions

Learning outcomes

- Know the factors that increase the rate of a reaction
- Know the factors that affect the equilibrium in a reversible reaction
- Understand the meaning of the term electrolysis
- Understand oxidation and reduction in terms of loss or gain of electrons or loss or gain of oxygen

Resources

Student Book page 302

Approach

This activity is only necessary if you are teaching the whole topic. Start by reviewing students' understanding of previous topics of work:

- Redox reactions as indicated by addition or loss of oxygen
- Factors affecting the rate of a reaction, especially temperature and the use of a catalyst – how do these factors affect the rate of a reaction?
- Enthalpy changes in exothermic and endothermic reactions – the convention of using ΔH with a positive or negative sign
- Factors affecting the equilibrium position in a reversible reaction – temperature and pressure
- Redox reactions in terms of loss or gain of electrons
- Electrolysis – preferential discharge at the electrodes; half-equations

Explain that in the three main industrial processes in this topic, some of the ideas above underpin the choice of conditions used.

Activity C5d.2 The manufacture of ammonia

Learning objectives

- To study the Haber process for the manufacture of ammonia
- To study the use of ammonia in the manufacture of nitric acid
- To consider some of the important uses of ammonia

Learning outcomes

- Be able to describe how ammonia is manufactured from nitrogen and hydrogen in the Haber process
- Understand that the nitrogen is obtained from the air and the hydrogen from natural gas or the cracking of hydrocarbons
- Be able to describe the conditions used in the Haber process:
  a temperature of 450 °C, a pressure of 200 atmospheres, an iron catalyst
  the ammonia is liquefied and the nitrogen and hydrogen are recycled
- Understand that the choice of conditions in the Haber process provides a balance between maximising the yield of ammonia and the rate of the reaction
- Be able to describe that ammonia is used in the manufacture of nitric acid and fertilisers

Common misconceptions

A catalyst changes not only the rate of a reaction but the yield as well.

Resources

Student Book pages 302–304

Approach

Explain why ammonia is such an important chemical, emphasising its use in the manufacture of fertilisers, nitric acid and cleaning agents in particular. Consider the issue of getting the best balance of rate, equilibrium position and cost.

Refer to Fig. 5.36 on page 303 of the Student Book. Emphasise that the catalyst does not affect the equilibrium position, as it will increase the rates of both forward and backward reactions to the same extent. Students will have met this compromise in Topic 4d *Equilibria* in the Science in Context feature on page 256 of the Student Book.

Consider the oxidation of ammonia in the first stage of the manufacture of nitric acid. What has been oxidised; what has been reduced? The oxidation of nitrogen monoxide in the presence of water to form nitric acid is one of the reactions that produces nitric acid in the atmosphere, causing acid rain.

Answers

Page 304

1. a) Nitrogen is obtained from the air.
   b) Hydrogen is obtained from natural gas or the cracking of hydrocarbons.
   c) The forward reaction (exothermic) is favoured by a low temperature.
   d) Iron is the catalyst.
   e) The catalyst has no effect on the yield of ammonia (just the rate of the reaction).
2. In ammonium nitrate %N = 28/80 × 100 = 35%. In ammonium chloride %N = 14/53.5 × 100 = 26.2%

Activity C5d.3 The manufacture of sulfuric acid

Learning objectives

- To study the reactions of the contact process used in the manufacture of sulfuric acid
- To consider some of the important uses of sulfuric acid

Learning outcomes

- Be able to describe how sulfuric acid is made in the contact process, starting from sulfur and air, but the main reaction involves the oxidation of sulfur dioxide to sulfur trioxide
- Know that the essential conditions in the contact process include:
  a temperature of 450 °C
  a pressure of about 2 atmospheres
  a catalyst of vanadium(V) oxide
- Be able to describe the use of sulfuric acid in the manufacture of detergents, fertilisers and paints

Resources

Student Book pages 304–306

Approach

Explain why sulfuric acid is such an important chemical, emphasising its use in the manufacture of detergents, fertilisers and paints in particular.

Refer to the separate stages in the manufacture (Student Book page 305), with the key reaction being the oxidation of sulfur dioxide to sulfur trioxide. Consider the issue of getting the best balance of rate, equilibrium position and cost.

Again, emphasise that the catalyst does not affect the equilibrium position, as it will increase the rates of both forward and backward reactions to the same extent. Students will have met this compromise in Topic 4d *Equilibria* in the Science in Context feature on page 256 of the Student Book, but it might be worth looking at this again.

Explain why it is so important to dissolve sulfur trioxide in sulfuric acid before adding water, rather than adding water directly to sulfur trioxide.

The time you spend on this activity will depend on how well students have understood the reasoning for the choice of reaction conditions in manufacturing ammonia.

Answers

Page 306

1. A higher temperature is used to give a suitable rate of reaction.
2. The catalyst is vanadium(V) oxide.
3. Increasing the pressure would be uneconomical because the yield is already very high at 98%.
4. Sulfuric acid reacts with ammonia to produce ammonium sulfate.

$$2NH_3(aq) + H_2SO_4(aq) \rightarrow (NH_4)_2SO_4(aq)$$

Activity C5d.4 The manufacture of sodium hydroxide and chlorine

Learning objectives

- To study the electrolysis of concentrated sodium chloride in a diaphragm cell
- To use half-equations to summarise the reactions taking place at the electrodes
- To consider some of the important uses of sodium hydroxide and chlorine

Learning outcomes

- Be able to describe how sodium hydroxide and chlorine are manufactured by the electrolysis of concentrated sodium chloride solution in a diaphragm cell
- Be able to write half-equations for the reactions taking place at the electrodes in the diaphragm cell
- Be able to describe the important uses of sodium hydroxide, including the manufacture of bleach, paper and soap
- Be able to describe the important uses of chlorine, including sterilising water supplies, and in the manufacture of bleach and hydrochloric acid

Resources

Student Book pages 307–310

Worksheet C1i.3b The electrolysis of sodium chloride solution

Files on CD–ROM: C1i.3b_worksheet, C1i.3b_tech_notes

Approach

Your students will already have studied the electrolysis of dilute and concentrated sodium chloride solution in Topic 1i *Electrolysis*. However, they will not have come across a diaphragm cell before and the reason it is needed (to keep the chlorine and sodium hydroxide apart).

Students could electrolyse sodium chloride solution using a simple cell, as indicated on Worksheet C1i.3b, as an introduction to the industrial process.

Emphasis will need to be placed on the ions present in the solution, including those from the water, the selective discharge at the cathode (linked to the reactivity series) and the half-equations that summarise the reactions that occur.

The uses of sodium hydroxide and chlorine are illustrated on page 308 of the Student Book.

Technician's notes

If you decide to carry out the practical activity, you will need the following resources for a class practical:

See Topic C1i and Activity Sheet C1i.3b.

Answers

Page 309

1. a) Na^+, Cl^-, H^+, OH^-

   b) Na^+ and H^+ ions are attracted to the cathode. The more reactive ion (less reactive atom) will be discharged, H^+.

   c) $2H^+(aq) + 2e^- \rightarrow H_2(g)$

2. Sodium hydroxide is used to manufacture bleach, paper and soap.

Page 310

1. Main points which can be included: ammonia NH_3; manufactured by Haber process $N_2(g) + 3H_2(g) \leftrightarrow 2NH_3(g)$; Nitrogen for this process obtained from the air. Obtained by fractional distillation – need to consider costs of obtaining this raw material.
   Sulfuric acid H_2SO_4: Manufactured by the contact process

   Step 1: sulfur burned in excess dry air to form sulphur dioxide; $S(s) + O_2(g) \rightleftharpoons SO_2(g)$

   Step 2: The gas mixture (sulphur dioxide and air) is filtered and then passed over a series of converters containing a catalyst where sulphur dioxide and oxygen from the air produce sulphur trioxide:
   $2SO_2(g) + O_2(g) \rightarrow 2SO_3(g)$

2. Main points which can be included:

   A catalyst is a substance that increases the rate of a chemical reaction and is chemically unchanged at the end of the reaction.

   Catalysts work by providing an alternative 'route' for the reaction which has a lower activation energy. The lower activation energy means that more of the collisions per second between particles will be effective. (These can be linked to iron and vanadium (V) oxide). See also Fig. 4.34.

   a) Haber process: Manufacture of ammonia

      $N_2(g) + 3H_2(g) \rightleftharpoons 2NH_3(g)$

      Exothermic reaction under normal conditions of temperature and pressure is slow.

      Finely divided iron is used as catalyst to increase the rate of reaction – large surface area as the catalyst is surface active (reaction takes place on the surface of the catalyst).

      Catalyst will increase the rate of both the forward and reverse reaction.

      Catalyst will enable the equilibrium position to be reached more rapidly.

   b) Contact process: Manufacture of sulfuric acid Sulphur dioxide and oxygen combine in contact with the catalyst vanadium(V) oxide (in a series of converters containing trays of catalyst) in an exothermic reaction to produce sulphur trioxide; $2SO_2(g) + O_2(g) \rightleftharpoons 2SO_3(g)$.

      Gas reactions take place on surface of catalyst – therefore need large surface area. Reaction needs to be dust free as this will reduce the effective surface area 'poisoning it' and limiting its efficiency (although vanadium(V) oxide is an efficient catalyst and is not easily poisoned).

Activity C5d.5 Consolidation and summary

Learning objectives

- To review the learning points of the topic
- To test understanding through answering questions

Learning outcomes

- Be familiar with the knowledge and understanding summarised in the End of Topic Checklist
- Be able to apply this knowledge and understanding by answering the End of Topic Questions

Resources

Student Book pages 309–313

Approach

You will only need this activity if you teach this topic as a whole. Recap the key learning points of the topic and focus on the Checklist and End of Topic Questions.

Refer to the Science in Context feature on page 309 of the Student Book. Ask students for their views on how manufacturers should present the hazards associated with the products they produce.

Answers

End of Topic Questions mark scheme

Question	Correct answer	Marks
1 a)	Nitrogen Hydrogen	1 mark 1 mark
1 b)	$N_2(g) + 3H_2(g) \leftrightarrow 2NH_3(g)$ (1 mark for formulae and state symbols; 1 mark for balancing; 1 mark for reversible reaction sign) Use correct sign for reversible reaction.	3 marks
1 c)	Completed table as shown	3 marks

Reaction conditions	Conditions chosen
Temperature	450 °C
Pressure	200 atmospheres
Catalyst	Iron

1 d)	At a lower temperature the rate of reaction is very low. The higher temperature used is a compromise; it increases the rate but reduces the yield.	1 mark 1 mark
1 e)	Cooling allows the ammonia formed to liquify. Removing the ammonia causes the equilibrium to move to the right so converting nitrogen and hydrogen into ammonia.	1 mark 1 mark

2 a)	$2SO_2(g) + O_2(g) \leftrightarrow 2SO_3(g)$ (1 mark for formulae and state symbols; 1 mark for balancing; 1 mark for reversible reaction sign.) Use correct sign for reversible reaction.	3 marks
2 b)	Complete table as shown below.	3 marks

Reaction conditions	Conditions chosen
Temperature	450 °C
Pressure	Normal pressure
Catalyst	Vanadium(V) oxide

2 c)	Sulfuric acid is used to make fertilisers/detergents/paints and dyes. Any two.	2 marks
3 a)	i) Chlorine is formed at the anode. ii) $2Cl^-(aq) \rightarrow Cl_2(g) + 2e^-$ (1 mark for symbols and state symbols; 1 mark for balancing)	1 mark 2 marks
3 b)	i) Hydrogen is formed. ii) $2H^+(aq) + 2e^- \rightarrow H_2(g)$ (1 mark for symbols and state symbols; 1 mark for balancing)	1 mark 2 marks
3 c)	The sodium hydroxide forms in the cell. From the Na^+ and OH^- ions that are not discharged.	1 mark 1 mark
	Total:	28 marks

Section 5: Exam-style questions mark scheme

Question	Correct answer	Marks
2 a)	The alkene homologous series	1 mark
2 b)	Monomer	1 mark
2 c)	Use Fig 5.26 but replace the Cl atoms with CH_3 units	1 mark 1 mark
2 d)	Again use Fig 26 for the repeat unit BUT replace Cl by CH_3 and only show a chain with 2 carbon atoms	1 mark 1 mark
2 e)	Addition polymer	1 mark
2 f)	Packaging/ ropes/ carpets (Any two)	2 marks
2 g)	They are non-biodegradable They are not easily broken down by bacteria in the soil	1 mark 1 mark
2 h)	i) It is a condensation polymer It is made from two different monomers	1 mark 1 mark
3 a)	A: haematite B: molten iron C: slag	1 mark 1 mark 1 mark
3 b)	i) $C(s) + O_2(g) \rightarrow CO_2(g)$ ii) $CO_2(g) + C(s) \rightarrow 2CO(g)$	1 mark 1 mark
3 c)	The limestone decomposes to form calcium oxide (CaO). The calcium oxide then combines with sand (silica) to form a slag.	1 mark 1 mark
3 d)	The iron(III) oxide has lost oxygen.	1 mark
3 e)	i) Aluminium is higher in the reactivity series than carbon. ii) Any two of the following use/property combinations: packaging (non-toxic); In aeroplanes (high strength to weight ratio – low density); electrical cables (good electrical conductivity); kitchen utensils (shiny appearance/non-toxic).	1 mark 1 mark 2 marks
4 a)	i) Bitumen ii) Gasoline iii) Bitumen	1 mark 1 mark 1 mark
4 b)	(Catalytic) cracking (1) The long-chain hydrocarbon is passed at high temperature (600–700 °C) over a catalyst of silica or alumina.	1 mark 1 mark 1 mark
4 c)	i) $2CH_4(g) + 3O_2(g) \rightarrow 2CO(g) + 4H_2O(l)$ ii) Carbon monoxide molecules attach themselves to the haemoglobin in the blood and this prevents the haemoglobin from transporting oxygen.	2 marks 1 mark 1 mark

5 a)	Reversible reaction	1 mark
5 b)	The reaction is exothermic	1 mark
5 c)	A low temperature will give a high yield A high temperature will give a good rate of reaction 450°C provides an optimum balance of reasonable yield at reasonable rate	1 mark 1 mark 1 mark
5 d)	i) A catalyst provides an alternative pathway for the reaction With a lower activation energy ii) Vanadium(V) oxide is the catalyst The reaction between sulfur trioxide and water is highly exothermic The reaction with oleum is much less violent and so is safer.	1 mark 1 mark 1 mark 1 mark 1 mark

Scheme of work

Activity	Learning outcomes and objectives	Specification reference	Student Book pages	Practical activity
C1a.1 Orientation	Learning objectives • To classify substances as solid, liquid or gas • To be familiar with the properties of typical solids, liquids and gases Learning outcomes • Know the three states of matter and their shorthand notation (s, l, g) • Know the typical properties of the three states of matter in terms of volume and shape	1.1	10–11	Demonstration of varying properties (hardness, viscosity, colour)
C1a.2 Kinetic theory of matter	Learning objective • To introduce the kinetic theory of matter Learning outcomes • Know the arrangement of particles in solids, liquids and gases • Understand the movement and energy of the particles in solids, liquids and gases	1.2	11–12	
C1a.3 Changes of state	Learning objectives • To reinforce the names of the main processes involved in changes of state • To use the kinetic theory of matter to explain the changes that occur Learning outcomes • Be able to describe how the three states of matter can be interconverted and know the names for these processes • Understand how the interconversions of melting, boiling, evaporation, freezing and condensation take place by describing changes in the arrangement, movement and energy of the particles involved	1.3	12–14	
C1a.4 Consolidation and summary	Learning objectives • To review the key learning points of the topic • To test understanding through answering questions Learning outcomes • Be familiar with the knowledge and understanding summarised in the End of Topic Checklist • Be able to apply this knowledge and understanding by answering the End of Topic Questions	1.1–1.3	16–17	
C1b.1 Orientation	Learning objectives • To review knowledge of atoms, molecules and compounds • To review knowledge of the separation of mixtures	1.4	18, 23–25	Demonstrations of , fractional distillation, filtration, crystallisation and

Activity	Learning outcomes and objectives	Specification reference	Student Book pages	Practical activity
	Learning outcomes • Know that an atom is the smallest part of an element that can exist on its own • Know that mixtures can be separated in a number of different ways			chromatography
C1b.2 Diffusion	Learning objectives • To observe and discuss a range of diffusion experiments • To observe and discuss Brownian motion Learning outcomes • Be able to describe simple diffusion experiments including dissolving a solid in a liquid and the mixing of gases • Be able to describe Brownian motion and explain the causes of the motion for pollen grains on water and smoke particles in air	1.4	18–20	Diffusion of potassium manganate(VII) in water; mixing carbon dioxide and air with lime water Demonstration of hydrogen chloride and ammonia, Brownian motion in a smoke cell
C1b.3 Elements, atoms, molecules, compounds and mixtures	Learning objectives • To provide definitions for the terms • To provide opportunities to test understanding of the terms Learning outcomes • Be able to define the terms element, atom, molecule, compound and mixture • Understand the differences between elements, compounds and mixtures	1.5, 1.6	21–22	
C1b.4 Separation of mixtures	Learning objective • To reinforce understanding of the different methods of separating mixtures Learning outcomes • Be able to describe techniques for separating mixtures including: ○ Simple distillation to separate a solid and liquid from a solution ○ Fractional distillation to separate liquids with different boiling points ○ Filtration to separate a solid from a liquid ○ Crystallisation to obtain a solid from a solution ○ Chromatography to separate different soluble solids from a solution	1.7, 1.8	23–25	Separation of mixtures by simple distillation and paper chromatography
C1b.5 Consolidation and summary	Learning objectives • To review the key learning points of the topic • To test understanding through answering questions Learning outcomes	1.4–1.8	26–27	

Activity	Learning outcomes and objectives	Specification reference	Student Book pages	Practical activity
	• Be familiar with the knowledge and understanding summarised in the End of Topic Checklist • Be able to apply this knowledge and understanding by answering the End of Topic Questions			
C1c.1 Orientation	Learning objectives • To revise students' knowledge about atoms • To review students' existing knowledge of atomic structure Learning outcomes • Know that atoms are the building blocks of all materials • Understand that atoms are made up of smaller sub-atomic particles which give the atom its characteristic properties	1.9	28	
C1c.2 Sub-atomic particles	Learning objectives • To introduce the structure of the atom • To show how atomic structure can be presented in terms of atom diagrams Learning outcomes • Know that an atom is made up of a central nucleus, composed of protons and neutrons, surrounded by electrons arranged in shells • Know the relative masses and charges of protons, neutrons and electrons • Understand the meaning of the terms atomic number, mass number and isotope	1.9– 1.11, 1.13	28–31	
C1c.3 Relative atomic mass and the mole	Learning objectives • To introduce the relative mass scale for comparing the mass of different atoms • To introduce the concept of a mole of atoms Learning outcomes • Know that the relative atomic mass of an atom is the relative mass of an atom on a scale where the mass of a carbon atom is 12 • Understand that the mole is a number equal to 6×10^{23} • Be able to perform simple calculations relating moles, mass and relative atomic mass • Be able to calculate the relative atomic mass of an element from the relative abundances of its isotopes	1.9, 1.12, 1.14	32–34	
C1c.4 Electron arrangements	Learning objectives • To explain how electrons are arranged in shells and how the number in the outer shell relates to the element's position in the Periodic Table • To introduce atom diagrams as a means of representing the atomic structure of atoms Learning outcomes • Understand that the Periodic Table is an arrangement of elements in order of their atomic numbers • Know that the arrangement of electrons in an atom can be worked out from its position in the Periodic Table	1.14, 1.15	34–37	

Activity	Learning outcomes and objectives	Specification reference	Student Book pages	Practical activity
	• Know that the number given to a group of elements in the Periodic Table is the same as the number of electrons in the outer electron shell of the atoms in the group			
C1c.5 Consolidation and summary	Learning objectives • To review the learning points of the topic • To test understanding through answering questions Learning outcomes • Be familiar with the knowledge and understanding summarised in the End of Topic Checklist • Be able to apply this knowledge and understanding by answering the End of Topic Questions	1.9–1.15	38–40	
C1d.1 Orientation	Learning objectives • To revise ideas on relative atomic masses and the mole • To map out the main purposes of the topic Learning outcomes • Know that the masses of different elements or atoms are measured on a relative scale • Understand what relative atomic mass (A_r) is • Understand that the mole is a number of atoms	1.16	41–42	Demonstration: reaction of magnesium ribbon
C1d.2 Relative formula masses	Learning objectives • To introduce the concept of formula mass and provide practice in calculating formula masses • To extend the mole concept to include compounds Learning outcomes • Know how to calculate relative formula masses from relative atomic masses • Understand the use of the term mole to represent the amount of a substance • Know that the Avogadro number of particles is 6×10^{23}	1.16–1.18	41–43	
C1d.3 Molar volumes of gases	Learning objectives • To introduce the concept of the molar volume of gases • To introduce reacting mass and volume calculations using balanced equations Learning outcomes • Know that the molar gas volume for any gas is 24 dm^3 (24 000 cm^3) at room temperature and pressure (RTP) • Be able to calculate quantities using relative atomic masses, relative formula masses and the mole	1.19, 1.20	44–46	
C1d.4 Consolidation and summary	Learning objectives • To review the learning points of the topic • To test understanding through answering questions	1.16–1.20	47–49	Calculating the molar volume of carbon dioxide

Activity	Learning outcomes and objectives	Specification reference	Student Book pages	Practical activity
	Learning outcomes • Be familiar with the knowledge and understanding summarised in the End of Topic Checklist • Be able to apply this knowledge and understanding by answering the End of Topic Questions			
C1e.1 Orientation	Learning objectives • To revise ideas on relative atomic masses, formula masses and the mole • To map out the main purposes of the topic Learning outcomes • Know how to work out a formula mass from atomic masses • Understand that a mole is a number that can be applied to atoms and molecules	Previous topic	50	Demonstration of combustion of magnesium ribbon in air
C1e.2 Finding the formula of magnesium oxide	Learning objectives • To allow students to undertake some exacting quantitative practical work • To show how a chemical formula can be worked out from combining masses of elements Learning outcomes • Understand how the formula of a simple compound such as a metal oxide can be worked out from experimental results • Be able to make accurate observations and measurements • Be able to analyse and interpret data to work out the formula of magnesium oxide • Be able to evaluate data and methods • Understand how the formulae of water and salts containing water of crystallisation can be worked out from experimental results	1.23	50–52	Working out the formula of magnesium oxide
C1e.3 Empirical and molecular formulae	Learning objectives • To introduce the empirical formula as a simple whole number ratio of the atoms in the compound • To emphasise the difference between and empirical formula and a molecular formula Learning outcomes • Be able to calculate empirical and molecular formulae from experimental data	1.23, 1.24	54–56	
C1e.4 Finding the formula of copper(II) oxide	Learning objectives • To provide another experimental method for determining the formula of a metal oxide • To encourage accurate quantitative work Learning outcomes • Understand how the formula of copper(II) oxide can be worked out from experimental results • Be able to make accurate observations and measurements • Be able to analyse and interpret data	1.24	51–53	Working out the formula of copper(II) oxide

Activity	Learning outcomes and objectives	Specification reference	Student Book pages	Practical activity
	• Be able to evaluate data and methods			
C1e.5 Writing chemical equations	Learning objectives • To write fully balanced chemical equations Learning outcomes • Be able to write balanced equations, including state symbols, for a range of chemical reactions	1.21	56–60	
C1e.6 Reacting masses	Learning objectives • To provide practise with reacting mass calculations • To introduce the concept of molarity as a means of denoting concentration • To introduce the concept of percentage yield Learning outcomes • Be able to use chemical equations to calculate reacting masses and the amounts of products formed • Understand how to carry out mole calculations using volumes and molar concentrations of solutions • Be able to calculate percentage yields in reactions by comparing actual amounts produced with those predicted by the chemical equation • Be able to make observations and measurements, analyse and interpret data and evaluate data and methods	1.25– 1.27	61–64	Obtaining zinc oxide from zinc carbonate
C1e.7 Consolidation and summary	Learning objectives • To review the learning points of the topic • To test understanding through answering questions Learning outcomes • Be familiar with the knowledge and understanding summarised in the End of Topic Checklist • Be able to apply this knowledge and understanding by answering the End of Topic Questions	1.21–1.27	65–67	
C1f.1 Orientation	Learning objectives • To revise ideas on atomic structure • To map out the main purposes of the topic Learning outcomes • Know the structure of an atom by interpreting atomic and mass numbers • Know the arrangement of electrons in shells for the first 20 elements	1.28	68	
C1f.2 The formation of ions	Learning objectives • To illustrate how metal atoms lose electrons and non-metal atoms gain electrons • To draw dot and cross diagrams to illustrate how atoms combine together Learning outcomes	1.29–1.32	68–71	

Activity	Learning outcomes and objectives	Specification reference	Student Book pages	Practical activity
	• Be able to describe the formation of ions by the gain or loss of electrons • Understand that oxidation is the loss of electrons and reduction is the gain of electrons • Know the charges of a range of common ions • Be able to deduce the charge on an ion from the electronic structure of the atom from which it is formed • Be able to explain the formation of ionic compounds of metals from Groups 1, 2 and 3 and non-metals from Groups 5, 6 and 7 using dot and cross diagrams			
C1f.3 Properties of ionic compounds	Learning objectives • To illustrate the characteristic properties of ionic compounds. • To explain the effect of the size of ion charges on melting and boiling points. Learning outcomes • Know that in the solid state the ions of an ionic compound are arranged in a giant ionic 3-dimensional (3D) lattice structure held together by the attractions between oppositely charged ions • Understand that ionic bonding is a strong electrostatic attraction between oppositely charged ions • Understand that ionic compounds have high melting and boiling points because of strong electrostatic forces between oppositely charged ions • Understand how the size of the charges in an ionic compound will determine the relative size of its melting and boiling points • Be able to draw a simple diagram to show the arrangement of sodium and chloride ions in a crystal of sodium chloride	1.33–1.37	71–73	
C1f.4 Consolidation and summary	Learning objectives • To review the learning points of the topic • To test understanding through answering questions Learning outcomes • Be familiar with the knowledge and understanding summarised in the End of Topic Checklist • Be able to apply this knowledge and understanding by answering the End of Topic Questions	1.28–1.37	74–75	
C1g.1 Orientation	Learning objectives • To revise ideas on atomic structure and ionic bonding • To map out the main purposes of the topic Learning outcomes • Know the structure of an atom from its atomic and mass numbers • Know the arrangement of electrons in shells for the first 20 elements	1.37	76	
C1g.2 The formation	Learning objectives • To illustrate how non-metals share electrons	1.38–1.39	76–79	

Activity	Learning outcomes and objectives	Specification reference	Student Book pages	Practical activity
of covalent compounds	• To draw dot and cross diagrams to show how non-metal atoms combine together Learning outcomes • Know that a covalent bond is formed when electrons are shared between the atoms of two non-metals • Understand that covalent bonding is a result of a strong attraction between the bonding pair of electrons and the nuclei of the atoms involved in the bond • Be able to explain the type of covalent bonding using dot and cross diagrams for the following: hydrogen, chlorine, hydrogen chloride, water, methane, ammonia, ethane, oxygen, nitrogen, carbon dioxide and ethene			
C1g.3 Properties of covalent compounds	Learning objectives • To illustrate the characteristic properties of covalent compounds • To provide an explanation for these properties in terms of the strength of forces between the molecules Learning outcomes • Know that simple molecular substances are gases, liquids or solids with low melting points • Be able to explain the properties of simple molecular substances in terms of the relatively weak forces between the molecules	1.40–1.42	80	
C1g.4 Diamond and graphite	Learning objectives • To illustrate the nature of the covalent bonding in diamond and graphite and to consider their differing structures • To use the structures to explain the characteristic properties of covalent molecules Learning outcomes • Know that giant covalent structures contain a lattice of strong covalent bonds • Be able to draw simple diagrams showing the arrangement of carbon atoms in diamond and graphite • Know the uses of diamond (as a cutting edge) and graphite (as a lubricant) and can explain these uses in terms of the structures of the two substances	1.43–1.44	80–82	
C1g.5 Consolidation and summary	Learning objectives • To review the learning points of the topic • To test understanding through answering questions Learning outcomes • Be familiar with the knowledge and understanding summarised in the End of Topic Checklist • Be able to apply this knowledge and understanding by answering End of Topic Questions	1.37–1.44	83–84	
C1h.1 Orientation	Learning objectives • To summarise knowledge and understanding of atomic structure and ionic bonding • To map out the main purposes of the topic	1.45	85	

Activity	Learning outcomes and objectives	Specification reference	Student Book pages	Practical activity
	Learning outcomes • Know that an atom is made up of a positively charged nucleus containing protons and neutrons (except hydrogen) with electrons in shells around the nucleus • Know that metal atoms lose electrons when they form ionic compounds			
C1h.2 Metallic bonding	Learning objectives • To introduce a model illustrating the structure of a metal • To use the model to explain the typical physical properties of metals • To undertake some research of alloys Learning outcomes • Be able to describe a metal as a giant structure of positive ions surrounded by a 'sea' of delocalised electrons • Be able to explain the electrical conductivity of metals in terms of the sea of delocalised electrons • Be able to explain the malleability of a metal in terms of its structure and bonding • Know that an alloy is a mixture of a metal and one or more other elements	1.45–1.46	85–87	
C1h.3 Consolidation and summary	Learning objectives • To review the learning points of the topic • To test understanding through answering questions Learning outcomes • Be familiar with the knowledge and understanding summarised in the End of Topic Checklist • Be able to apply this knowledge and understanding by answering the End of Topic Questions	1.45–1.46	88–89	
C1i.1 Orientation	Learning objectives • To revise ideas on chemical bonding (especially ionic bonding) and metallic bonding • To introduce the terminology associated with the process of electrolysis • To map out the main purposes of the topic Learning outcomes • Know that the anode is a positive electrode • Know that the cathode is a negative electrode • Understand that an electric current is a flow of electrons in a conductor or ions in an electrolyte • Know that electrolysis is the breaking down of a compound by the passage of electricity (an electric current) • Know that an electrolyte is a substance that, when molten or dissolved in water, allows an electric current to pass through it	1.47–1.49	90–92	Demonstration: conduction through metal and sodium chloride solution
C1i.2	Learning objectives	1.50–1.52	92–93	Demonstration: molten lead(II)

Activity	Learning outcomes and objectives	Specification reference	Student Book pages	Practical activity
Electrolysis of molten electrolytes	• To demonstrate the electrolysis of a molten electrolyte, lead(II) bromide • To emphasise the requirement for ions to be able to move if electrolysis is to take place • To introduce half-equations Learning outcomes • Know that for a substance to act as an electrolyte the ions must be free to move • Be able to describe the electrolysis of molten substances such as lead(II) bromide • Be able to write half-equations to represent the reactions that occur at the electrodes			bromide conducting electricity

C1i.3
Electrolysis of aqueous electrolytes
	Learning objectives
	• To allow the students to identify electrolytes from a range of liquids and solutions
	• To provide an opportunity to use knowledge of ionic and covalent bonding to make predictions about potential electrolytes
	Learning outcomes
	• Be able to describe the electrolysis of the following electrolytes and predict the products formed:
	  ○ aqueous sodium chloride solution
	  ○ aqueous copper(II) sulfate and dilute sulfuric acid
	• Be able to write half-equations to represent the reactions that occur at the electrodes
		1.53–1.54	93–97	Identifying and investigating electrolytes
				The electrolysis of sodium chloride solution

C1i.4 Quantitative electrolysis	Learning objectives • To introduce the link between half equations and the mole of electrons (the faraday) Learning outcomes • Be able to write half-equations to represent the reactions that occur at electrodes • Know that a faraday is a mole of electrons • Understand how to calculate the amounts of products formed in the electrolysis of molten compounds and aqueous solutions from the half equations involved	1.54–1.56	98	To find out how many faradays of electricity are needed to produce 1 mole of copper
C1i.5 Consolidation and summary	Learning objectives • To review the learning points of the topic • To test understanding through answering questions Learning outcomes • Be familiar with the knowledge and understanding summarised in the End of Topic Checklist • Be able to apply this knowledge and understanding by answering the End of Topic Questions	1.47–1.56	99–101	
C2a.1 Orientation	Learning objectives • To revise understanding of what an element is • To revise knowledge of atomic structure	2.1	108	Demonstration: different elements

Activity	Learning outcomes and objectives	Specification reference	Student Book pages	Practical activity
	Learning outcomes • Know that an element is a substance that cannot be broken down into anything simpler by chemical means • Known how atomic structure is related to the atomic number • Know how electrons are arranged in shells around the nucleus			
C2a.2 The arrangement of the elements	Learning objectives • To look in more detail at the arrangement of elements in the Periodic Table • To emphasise the positions of metals/non-metals/metalloids; trends in reactivity; electron arrangements and ion charges Learning outcomes • Know that metals are found on the left-hand side and in the middle of the Periodic Table • Know that non-metals are found on the right-hand side of the Periodic Table • Understand that the chemical properties of an element are largely governed by the number of electrons in the outermost electron shell • Know that in groups of metals reactivity increases down the group whereas in groups of non-metals reactivity decreases down the group • Understand that the noble gases in Group 0 are very unreactive because they have full outer shells or 8 electrons in their outer shells	2.2–2.5	108–110	
C2a.3 Characteristic properties of metals and non-metals	Learning objectives • To explore the physical and chemical properties of a range of metals and non-metals • To identify the characteristic properties of metals and non-metals Learning outcomes • Be able to explain that metals conduct electricity and form basic oxides • Be able to explain that non-metals are generally non-conductors of electricity and usually form acidic oxides	2.3	110–113	Physical properties of metals and non-metals Investigating oxides
C2a.4 Consolidation and summary	Learning objectives • To review the learning points of the topic • To test understanding through answering questions Learning outcomes • Be familiar with the knowledge and understanding summarised in the End of Topic Checklist • Be able to apply this knowledge and understanding by answering End of Topic Questions	2.1–2.5	114–115	
C2b.1 Orientation	Learning objectives • To revise the nature of the grouping of elements in the Periodic Table • To establish Group 1 as a 'family' of very reactive metals Learning outcomes	2.6	116–117	Demonstration: lithium, sodium and potassium

Activity	Learning outcomes and objectives	Specification reference	Student Book pages	Practical activity
	• Know that lithium, sodium and potassium are soft, reactive metals and stored in oil			
C2b.2 Reactions of the Group 1 elements	Learning objectives • To demonstrate the reactions of lithium, sodium and potassium with water • To write balanced chemical equations to represent these reactions Learning outcomes • Be able to describe the reactions of lithium, sodium and potassium with water • Know that in their reactions with water an alkali and hydrogen are produced • Know that the elements become increasingly reactive as you go down the group, from lithium to francium	2.6–2.7	118–121	Demonstration: reactions of lithium, sodium and potassium with water
C2b.3 Consolidation and summary	Learning objectives • To provide an explanation for the trend in reactivity of the Group 1 elements • To review the learning points of the topic • To test understanding through answering questions Learning outcomes • Be able to explain that the reactivity of the elements depends on the distance between the outer electron and the nucleus; the further the distance of the outer electron from the nucleus, the more reactive the metal • Be familiar with the knowledge and understanding summarised in the End of Topic Checklist • Be able to apply this knowledge and understanding by answering End of Topic Questions	2.6–2.8	122–123	
C2c.1 Orientation	Learning objectives • To review the significance of the positioning of the Group 7 elements in the Periodic Table and the trend in reactivity for a group of non-metals • To study the appearances of the Group 7 elements Learning outcomes • Know the colours and physical states of chlorine, bromine and iodine at room temperature • Know the relative reactivity of the Group 7 elements • Be able to predict the likely properties of fluorine and astatine	2.9–2.10, 2.13	124–125	Demonstration: chlorine, bromine and iodine
C2c.2 Displacement reactions of the halogens	Learning objectives • To consider the nature of the displacement reactions involving the halogen elements • To introduce or revise the concept of redox reactions Learning outcomes • Be able to describe and explain displacement reactions involving the halogens and solutions of their salts • Understand that displacement reactions are redox reactions and involve oxidation (loss of electrons) and reduction (gain of electrons)	2.14–2.15	126–129	Displacement reactions of halogens

Activity	Learning outcomes and objectives	Specification reference	Student Book pages	Practical activity
C2c.3 Hydrogen chloride	Learning objectives • To contrast hydrochloric acid (a solution of hydrogen chloride in water) with a solution of hydrogen chloride in methylbenzene Learning outcomes • Understand that hydrogen chloride dissolves in water to form hydrochloric acid which dissociates into hydrogen ($H^+(aq)$) ions and chloride ($Cl^-(aq)$) ions • Understand that hydrogen chloride in other solvents, such as methylbenzene, does not dissociate into ions and so shows no acidic properties	2.11–2.12	129–131	Demonstration: conductivity of hydrogen chloride
C2c.4 Consolidation and summary	Learning objectives • To review the learning points of the topic • To test understanding through answering questions Learning outcomes • Be familiar with the knowledge and understanding summarised in the End of Topic Checklist • Be able to apply this knowledge and understanding by answering End of Topic Questions	2.9–2.15	132–133	
C2d.1 Orientation	Learning objectives • To revise the concept of basic and acidic oxides • To identify the gases which are the main components of the air Learning outcomes • Know that a basic oxide is the oxide of a metal and an acidic oxide is the oxide of a non-metal • Know that the air contains approximately 78% nitrogen, 21% oxygen, 0.9% argon and 0.04% carbon dioxide • Be able to describe the reactions of oxygen with magnesium, carbon and sulfur and write equations for these reactions	2.16–2.17	134–135	Demonstration: reaction of oxygen with magnesium, phosphorus, sulfur and carbon
C2d.2 Determining the percentage composition of oxygen in the air	Learning objectives • To demonstrate an experiment to determine the percentage of oxygen in the air • To consider other methods of determining the percentage of oxygen in the air Learning outcomes • Be able to explain how the reactions of copper, iron and phosphorus with air can be used to determine the percentage by volume of oxygen in the air • Know that oxygen accounts for approximately one-fifth of the air	2.17	135–136, 138	Demonstration: formation of copper oxide to demonstrate amount of oxygen in the air
C2d.3 The preparation and properties	Learning objectives • To prepare oxygen and identify its presence using a glowing splint Learning outcomes • Be able to describe the laboratory preparation of oxygen from hydrogen peroxide using manganese(IV)	2.18–2.19	136–137	The preparation and properties of oxygen

Activity	Learning outcomes and objectives	Specification reference	Student Book pages	Practical activity
of oxygen	oxide as a catalyst • Know that the relighting of a glowing splint is used as a simple identification test for oxygen			
C2d.4 The preparation and properties of carbon dioxide	Learning objectives • To prepare and identify carbon dioxide • To relate the uses of carbon dioxide to its properties Learning outcomes • Be able to describe the preparation of carbon dioxide from calcium carbonate and dilute hydrochloric acid • Be able to describe the formation of carbon dioxide from the thermal decomposition of copper(II) carbonate • Be able to explain that carbon dioxide is used as the gas in fizzy drinks as it dissolves in water • Be able to explain that carbon dioxide is used in fire extinguishers as it is denser than air	2.20–2.23	138–140	The preparation and properties of carbon dioxide
C2d.5 Carbon dioxide as a greenhouse gas	Learning objectives • To consider the potential detrimental effects of the formation of carbon dioxide from the combustion of fuels. Learning outcomes • Know that carbon dioxide is a greenhouse gas and may contribute to climate change.	2.24–2.25	140–142	
C2d.6 Consolidation and summary	Learning objectives • To review the learning points of the topic To test understanding through answering questions Learning outcomes • Be familiar with the knowledge and understanding summarised in the End of Topic Checklist • Be able to apply this knowledge and understanding by answering End of Topic Questions	2.17–2.25	143–144	
C2e.1 Orientation	Learning objectives • To identify the position of hydrogen in the Periodic Table • To revise the atomic structure of hydrogen Learning outcomes • Know that hydrogen is not placed in a group in the Periodic Table • Know the atomic structure of hydrogen	2.26	145	
C2e.2 The preparation of hydrogen	Learning objectives • To prepare hydrogen and identify its presence with a lighted splint • To investigate the different reactivities of metals with dilute acids Learning outcomes • Be able to describe how hydrochloric acid reacts with magnesium, aluminium, zinc and iron to produce	2.26	145–147	The preparation of hydrogen

Activity	Learning outcomes and objectives	Specification reference	Student Book pages	Practical activity
	hydrogen and a metal chloride • Be able to describe how sulfuric acid reacts with magnesium, aluminium, zinc and iron to produce hydrogen and a metal sulfate • Be able to describe a simple test to identify hydrogen			
C2e.3 Water	Learning objectives • To demonstrate the formation of water when hydrogen burns • To know how to identify the presence of water and verify that the water is pure Learning outcomes • Be able to describe how hydrogen burns in air or oxygen to form water • Be able to describe how anhydrous copper(II) sulfate can be used to identify the presence of water and that the anhydrous copper(II) sulfate changes colour from white to blue when water is present • Be able to describe a test to show whether water is pure or not	2.27–2.29	146–148	Demonstration: burning hydrogen, action of heat on hydrated copper(II) sulfate crystals
C2e.4 Consolidation and summary	Learning objectives • To review the learning points of the topic • To test understanding through answering questions Learning outcomes • Be familiar with the knowledge and understanding summarised in the End of Topic Checklist • Be able to apply this knowledge and understanding by answering the End of Topic Questions in the Student Book	2.26–2.29	149–150	
C2f.1 Orientation	Learning objectives • To recap on existing knowledge of the differing reactivities of metals with water and dilute hydrochloric or sulfuric acids Learning outcomes • Understand that metals can be arranged in a reactivity series: potassium, sodium, lithium, calcium, magnesium, aluminium, zinc, iron, copper, silver and gold • Be able to describe how the reactions of metals with water and dilute acids can be used to deduce an order of reactivity	2.30–2.31	151–152, 156	Demonstration: reactions of sodium with water, magnesium with steam
C2f.2 Displacement reactions	Learning objectives • To introduce or reinforce ideas on competition or displacement reactions • To introduce the characteristics of a redox reaction Learning outcomes • Understand how displacement reactions between metals and metal oxides and between metals and metal salts in aqueous solution can be used to deduce the order of reactivity of metals	2.31–2.34	152–154	Metals and displacement reactions Demonstration: displacement reaction of magnesium powder with copper(II) oxide

Activity	Learning outcomes and objectives	Specification reference	Student Book pages	Practical activity
	• Understand that oxidation is the addition of oxygen • Understand that reduction is the removal of oxygen • Understand that an oxidising agent will oxidise another element or compound but will itself be reduced • Understand that a reducing agent will reduce another element or compound but will itself be oxidised • Understand the term redox			
C2f.3 Limiting corrosion	Learning objectives • To consider the relative merits of different ways of preventing iron from rusting • To undertake a devise and plan investigation Learning outcomes • Be able to describe that air (oxygen) and water must be present for iron to rust • Be able to describe how rusting can be prevented by using grease, oil, plastic or by galvanising • Understand the sacrificial protection of iron in terms of the reactivity series	2.35–2.37	155–157	Preventing corrosion
C2f.4 Consolidation and summary	Learning objectives • To review the learning points of the topic • To test understanding through answering questions Learning outcomes • Be familiar with the knowledge and understanding summarised in the Student Book Checklist • Be able to apply this knowledge and understanding by answering the End of Topic Questions	2.30–2.37	158–159	
C2g.1 Orientation	Learning objectives • To recap on existing knowledge of ions and identification tests for gases Learning outcomes • Know that metals form positive ions which are called cations • Know that non-metals form negative ions which are called anions • Know the identification tests for hydrogen, oxygen and carbon dioxide	2.38–2.39	160	
C2g.2 Identifying cations	Learning objectives • To detect the presence of certain metal ions (cations) using a flame test • To introduce the test for the ammonium ion • To explore the use of sodium hydroxide solution to detect metal ions in solution Learning outcomes • Be able to describe a flame test using a nichrome wire, concentrated hydrochloric acid and a sample of a solid • Know the flame colours produced by the following metal ions: Li^+; Na^+; K^+ and Ca^{2+}	2.38	160–162	Using a flame test to identify metal ions Identifying metal ions in solution Demonstration: identifying the ammonium ion

Activity	Learning outcomes and objectives	Specification reference	Student Book pages	Practical activity
	• Be able to describe how to test for the ammonium ion, NH_4^+, using sodium hydroxide and know what indicates a positive result in each case • Be able to describe tests for Cu^{2+}, Fe^{2+} and Fe^{3+} ions using sodium hydroxide solution and know what indicates a positive result in each case			
C2g.3 Identifying anions	Learning objectives • To identify the presence of the carbonate ion • To identify certain non-metal ions in solution Learning outcomes • Be able to describe a test for CO_3^{2-} ions using dilute hydrochloric acid and know what indicates a positive result • Be able to describe a test for Cl^-, Br^- and I^- ions using dilute nitric acid and silver nitrate solution and know the colours of the different precipitates • Be able to describe a test for SO_4^{2-} ions using dilute hydrochloric acid and barium chloride solution and know what indicates a positive result	2.39	162–164	Identifying non-metal ions in solution
C2g.4 Identifying gases	Learning objectives • To revise some simple identification tests for common gases Learning outcomes • Be able to describe tests for the gases hydrogen, oxygen, carbon dioxide, ammonia and chlorine and know what indicates a positive result in each case	2.40	164–166	Identifying gases
C2g.5 Consolidation and summary	Learning objectives • To review the learning points of the topic • To test understanding through answering questions Learning outcomes • Be familiar with the knowledge and understanding summarised in the End of Topic Checklist • Be able to apply this knowledge and understanding by answering End of Topic Questions	2.38–2.40	167–168	
C3a.1 Orientation	Learning objectives • To revise the nature of covalent bonding • To introduce the concept of a homologous series with alkanes as an example Learning outcomes • Know that a homologous series is a group of organic compounds with the same general formula, similar chemical properties and physical properties which change gradually from one member to the next • Know that alkanes have a general formula of C_nH_{2n+2} • Know that a hydrocarbon is a compound containing carbon and hydrogen only	3.1	176	Demonstration: examples of the alkanes

Activity	Learning outcomes and objectives	Specification reference	Student Book pages	Practical activity
C3a.2 The structure of alkanes	Learning objectives • To look in more detail at the structure of alkanes, including non-straight chain isomers • To introduce the concept of a saturated/unsaturated hydrocarbon Learning outcomes • Be able to name the first five straight-chain members of the alkane homologous series (methane, ethane, propane, butane and pentane) • Be able to draw displayed formulae for the first five members of the alkane homologous series • Know that isomerism is the existence of compounds with the same molecular formulae but different structural formulae • Know that a saturated hydrocarbon contains only carbon to carbon single bonds (whereas an unsaturated hydrocarbon contains C=C double bonds)	3.1	176–180	
C3a.3 The chemical properties of alkanes	Learning objectives • To study the combustion reactions of alkanes and the nature of the products formed • To study the substitution reactions of alkanes and the nature of the products formed Learning outcomes • Know that when alkanes burn in a plentiful supply of air or oxygen they form carbon dioxide and water • Know that in a limited supply of air or oxygen (incomplete combustion) alkanes can burn to form carbon and carbon monoxide • Know that methane will react with bromine in the presence of UV light to form bromomethane (CH_3Br)	3.1, 3.4–3.5	181–182	
C3a.4 Consolidation and summary	Learning objectives • To review the learning points of the topic • To test understanding through answering questions Learning outcomes • Be familiar with the knowledge and understanding summarised in the End of Topic Checklist • Be able to apply this knowledge and understanding by answering the End of Topic Questions	3.1	182–184	
C3b.1 Orientation	Learning objectives • To recap knowledge of covalent bonding and the alkane homologous series • To introduce the names and displayed formulae for simple alkenes Learning outcomes • Know that alkenes have a general formula of C_nH_{2n} • Know that unsaturated hydrocarbons contain C=C double bonds • Be able to name the first three straight chain members of the alkene homologous series (ethene, propene, butene)	3.2–3.3	185	

Activity	Learning outcomes and objectives	Specification reference	Student Book pages	Practical activity
	• Be able to draw displayed formulae for the first three members of the alkene homologous series, including the isomers of butane			
C3b.2 The chemical properties of alkenes	Learning objectives • To look at the combustion of alkenes as an homologous series of hydrocarbons • To introduce the reaction of an alkene with bromine as an addition reaction Learning outcomes • Know that alkenes are hydrocarbons, so will burn in a plentiful supply of air to form carbon dioxide and water • Understand what an addition reaction with an alkene involves • Be able to describe the reaction of an alkene with bromine water and know that this reaction can be used to identify alkenes	3.4–3.8	185–188	Demonstration: reaction of alkene with bromine water
C3b.3 Consolidation and summary	Learning objectives • To review the learning points of the topic • To test understanding through answering questions Learning outcomes • Be familiar with the knowledge and understanding summarised in the End of Topic Checklist • Be able to apply this knowledge and understanding by answering the End of Topic Questions	3.2–3.8	187–190	
C3c.1 Orientation	Learning objectives • To introduce alcohols as another homologous series of organic compounds • To consider the covalent bonding in the alcohols Learning outcomes • Know that alcohols have a general formula of $C_nH_{2n+1}OH$ • Know that ethanol is the commonest alcohol	3.9–3.12	191–193	
C3c.2 The manufacture of ethanol	Learning objectives • To introduce two methods of manufacturing ethanol • To consider the factors that influence the choice of method used Learning outcomes • Be able to describe the fermentation process used in the manufacture of ethanol from sugar, including the importance of temperature and the use of yeast • Be able to describe the manufacture of ethanol using ethene and steam passed over a catalyst of phosphoric acid, with a temperature of 300 °C and pressure of 70 atmospheres • Be able to evaluate the factors relevant to the choice of method used in the manufacture of ethanol, including the relative availability of sugar cane and crude oil	3.9–3.11	193–196	Demonstration: fermentation

Activity	Learning outcomes and objectives	Specification reference	Student Book pages	Practical activity
C3c.3 Reactions of ethanol	Learning objectives • To consider the advantages of using ethanol as a fuel • To introduce the preparation of ethene by the dehydration of ethanol Learning outcomes • Understand the advantages of using ethanol as a fuel • Be able to describe the dehydration of ethanol to ethene using aluminium oxide	3.12	196–198	The dehydration of ethanol Demonstration: ethanol in spirit burner
C3c.4 Consolidation and summary	Learning objectives • To review the learning points of the topic • To test understanding through answering questions Learning outcomes • Be familiar with the knowledge and understanding summarised in the End of Topic Checklist • Be able to apply this knowledge and understanding by answering End of Topic Questions	3.9–3.12	199–200	
C4a.1 Orientation	Learning objectives • To revise the use of indicators to classify substances as acidic, neutral or alkaline • To compare different indicators • To reinforce the use of universal indicator to identify substances which are strongly or weakly acid or alkaline Learning outcomes • Know how indicators such as litmus, phenolphthalein and methyl orange can be used to distinguish between acidic and alkaline solutions • Know the colours of each of the above indicators in acidic and alkaline solutions • Know how universal indicator can be used to test a solution • Understand how the pH scale from 0 to 14 can be used to classify solutions as strongly acidic, weakly acidic, neutral, weakly alkaline or strongly alkaline	4.1–4.3	206–207	Using indicators
C4a.2 Acids, bases and alkalis	Learning objectives • To emphasise the nature of acids, bases, alkalis and salts, including hydrated salts • To introduce the characteristics of acids and alkalis in terms of the ions they provide in solution Learning outcomes • Know that acids when in aqueous solution are a source of H^+ ions • Know that alkalis when in aqueous solution are a source of OH^- ions • Know that salts made from hydrochloric acid are called chlorides • Know that salts made from nitric acid are called nitrates • Know that salts made from sulfuric acid are called sulfates	4.4	207–209	

Activity	Learning outcomes and objectives	Specification reference	Student Book pages	Practical activity
C4a.3 Making soluble salts	Learning objectives • To provide some simples rules for deciding whether a salt is soluble or insoluble • To explore methods for preparing soluble salts, including the reaction between an acid and an alkali using the titration technique Learning outcomes • Be able to predict the products of reactions between dilute hydrochloric acid, nitric and sulfuric acids and metals, metal oxides and metal carbonates • Be able to predict whether a simple salt is soluble or insoluble in water • Be able to predict whether a carbonate is soluble or insoluble in water • Know how soluble salts can be prepared by adding a metal, base or carbonate to an acid • Know how a soluble salt can be prepared by titration using an acid and an alkali • Be able to describe how to carry out an acid – alkali titration • Understand how to calculate the concentration of an acid or alkali from the results of a titration experiment	4.5–4.7, 4.9	209–214, 216–217	Making a soluble salt (1) – using a titration method Making a soluble salt (2)
C4a.4 Making insoluble salts	Learning objectives • To introduce the preparation of insoluble salts using the precipitation method Learning outcomes • Be able to describe how to prepare an insoluble salt using a precipitation reaction	4.8		Making an insoluble salt
C4a.5 Consolidation and summary	Learning objectives • To review the learning points of the topic • To test understanding through answering questions Learning outcomes • Be familiar with the knowledge and understanding summarised in the End of Topic Checklist • Be able to apply this knowledge and understanding by answering End of Topic Questions	4.1–4.9	217–221	
C4b.1 Orientation	Learning objectives • To revise understanding of covalent bonding • To measure temperature changes in a range of reactions • To classify reactions into those that give out energy and those that take energy in Learning outcomes • Know the nature of the bonds present in fuels such as the alkanes • Know that chemical reactions in which heat energy is given out are described as exothermic • Know that chemical reactions in which heat energy is taken in are described as endothermic	4.10	222–223	Exothermic and endothermic reactions

Activity	Learning outcomes and objectives	Specification reference	Student Book pages	Practical activity
C4b.2 Measuring energy transfers (1)	Learning objectives • To calculate energy changes in a reaction using a polystyrene cup as a simple calorimeter • To calculate energy changes in terms of the molar quantities given in the equation Learning outcomes • Know how a polystyrene cup can be used to carry out simple calorimetric experiments to determine energy changes • Understand how to use the equation: energy transferred = mass of water (g) × 4.2 (J/g/°C) × rise in temperature (°C) • Understand that heat energy is called enthalpy • Know that the molar enthalpy change (ΔH) is the heat energy change when the molar quantities shown in the chemical equation react together	4.11–4.13	222–227	Measuring energy transfers (1) Designing a 'cold pack'
C4b.3 Measuring energy transfers (2)	Learning objectives • To measure energy changes during the combustion of a fuel • To calculate energy changes in terms of the molar quantities given in the equation Learning outcomes • Know how a metal calorimeter can be used to measure the energy change when a fuel is burnt • Understand how to use the equation: energy transferred = mass of water (g) × 4.2 (J/g/°C) × rise in temperature (°C) • Understand that heat energy is called enthalpy • Know that the molar enthalpy change (ΔH) is the heat energy change when the molar quantities shown in the chemical equation react together • Understand how to calculate the molar enthalpy change from the results of an experiment and information about molar quantities	4.11–4.16	222–227	Measuring energy transfers (2)
C4b.4 Energy level diagrams and bond breaking and bond forming	Learning objectives • To introduce energy level diagrams • To highlight the differences in energy level diagrams for exothermic and endothermic reactions • To show the significance of bond breaking and bond making in the determination of the molar enthalpy change (ΔH) for a reaction • To use bond energy values to calculate molar enthalpy changes in reactions Learning outcomes • Understand that ΔH is used to represent the molar enthalpy change for a reaction • Know that ΔH for an exothermic reaction is negative and that ΔH for an endothermic reaction is positive • Understand how energy level diagrams can be used to show the difference between exothermic and endothermic reactions	4.13–4.16	228–230	

Activity	Learning outcomes and objectives	Specification reference	Student Book pages	Practical activity
	• Understand that the breaking of bonds is endothermic and that the making of bonds is exothermic • Be able to use average bond energies to calculate the enthalpy change in a simple chemical reaction			
C4b.5 Consolidation and summary	Learning objectives • To review the learning points of the topic • To test understanding through answering questions Learning outcomes • Be familiar with the knowledge and understanding summarised in the End of Topic Checklist • Be able to apply this knowledge and understanding by answering the End of Topic Questions	4.10–4.16	231–235	
C4c.1 Orientation	Learning objectives • To revise the particle arrangements and movement in the three states of matter • To demonstrate that reactions occur at different rates • To introduce a simple collision theory Learning outcomes • Be able to explain how particles are arranged and move in solids, liquids and gases • Know the difference between time, speed and rate • Know in terms of a collision theory what must happen for two particles to react together	4.17	236–237	Demonstration: reactions that occur at different speeds
C4c.2 Monitoring the rate of reaction	Learning objectives • To demonstrate that the rate of a reaction usually decreases as the reaction takes place • To how some of the experimental methods that can be used to measure the changing rate of a reaction • To show how experimental results can be used to work out the rate of a reaction at a particular time Learning outcomes • Be able to demonstrate and describe techniques showing how the rate of a reaction can be monitored by either collecting the gas produced or measuring the loss in mass as the gas escapes • Know how to make observations and measurements in reactions where the change in rate is being measured • Understand how to analyse and interpret data to draw conclusions about the rate of a reaction • Be able to evaluate data and methods in the experiments used • Be able to explain why rate is affected by concentration using a simple collision theory model	4.18–4.20	238–240	Monitoring the rate of a reaction
C4c.3 Surface area of solids	Learning objectives • To show that the rate of a reaction involving a solid and a liquid/solution depends on the surface area of the solid • To show how the effect of surface area on rate can be measured experimentally	4.18–4.20	240–245	Reaction rate and surface area Demonstration: monitoring the rate of a reaction

Activity	Learning outcomes and objectives	Specification reference	Student Book pages	Practical activity
	• To explain the effect of surface area on the rate of a reaction using a simple collision model Learning outcomes • Know how to make observations and measurements in reactions where the change in rate is being measured • Understand how to analyse and interpret data to draw conclusions about the rate of a reaction • Be able to evaluate data and methods in the experiments used • Be able to explain why rate is affected by surface area using a simple collision theory model			
C4c.4 Concentration of solutions	Learning objectives • To show that the rate of a reaction depends on the concentration of solutions • To show how the 'initial' rate of a reaction can be measured with different concentrations of reactants • To explain the effect of concentration on the rate of a reaction using a simple collision model Learning outcomes • Be able to demonstrate and describe a technique for comparing the rates of a reaction with different concentrations of solutions • Know how to make observations and measurements in reactions where the change in rate is being measured • Understand how to analyse and interpret data to draw conclusions about the rate of a reaction • Be able to evaluate data and methods in the experiment used • Be able to explain why rate is affected by concentration using a simple collision theory model	4.18–4.20	240–245	Reaction rate and concentration
C4c.5 Temperature	Learning objectives • To show that the rate of a reaction depends on temperature • To show how the rate of a reaction can be measured at different temperatures • To explain the effect of temperature on the rate of a reaction using a simple collision model Learning outcomes • Be able to devise and plan an investigation to show how temperature affects the rate of a reaction • Know how to make observations and measurements in reactions where rate is being measured • Understand how to analyse and interpret data to draw conclusions about the rate of a reaction • Be able to evaluate data and methods in the experiment used • Be able to explain why rate is affected by temperature using a simple collision theory model	4.18–4.20	240–245	Reaction rate and temperature
C4c.6 Use of a catalyst	Learning objectives • To show that catalysts can increase the rate of a reaction • To demonstrate the specificity of catalysts	4.21	246–247	Catalysts

Activity	Learning outcomes and objectives	Specification reference	Student Book pages	Practical activity
	• To show how manganese(IV) oxide catalyses the decomposition of hydrogen peroxide • Explain how a catalyst increases the rate of a reaction Learning outcomes • Know how to make observations and measurements in reactions where rate is being measured • Understand how to analyse and interpret data to draw conclusions about the rate of a reaction • Be able to evaluate data and methods in the experiment used • Be able to explain why rate is affected by the presence of a catalyst			
C4c.7 Consolidation and summary	Learning objectives • To revise the key learning points of the topic • To test understanding through answering questions Learning outcomes • Be familiar with the knowledge and understanding summarised in the End of Topic Checklist • Be able to apply this knowledge and understanding by answering the End of Topic Questions	4.17–4.21	248–251	
C4d.1 Orientation	Learning objectives • To introduce the concept of a reversible reaction • To study the actions of heat on copper(II) sulfate crystals and ammonium chloride Learning outcomes • Know that a reversible reaction is one in which reactants form products and products can reform reactants • Know that some reactions, such as heating copper(II) sulfate or ammonium chloride are reversible • Know the symbol used in a chemical equation to show a reversible reaction • Be able to describe the changes that occur when copper(II) sulfate crystals are heated and then when water is added • Be able to describe the changes that occur when ammonium chloride is heated	4.22–4.23	252–254	Reversible reactions
C4d.2 Changing the position of equilibrium	Learning objectives • To introduce the concept of a dynamic equilibrium • To consider the effects of changing the conditions of a reaction on the position of the equilibrium Learning outcomes • Understand and be able to explain the concept of a dynamic equilibrium • Be able to predict the effects of the following changes on a reversible or equilibrium reaction: ○ changing the concentration of reactants or products ○ changing the pressure ○ changing the temperature • Know that a catalyst affects the rates of forwards and backward reactions to the same extent and so does	4.24–4.25	254–258	

Activity	Learning outcomes and objectives	Specification reference	Student Book pages	Practical activity
	not change the position of equilibrium			
C4d.3 Consolidation and summary	Learning objectives • To review the learning points of the topic • To test understanding through answering questions Learning outcomes • Be familiar with the knowledge and understanding summarised in the End of Topic Checklist • Be able to apply this knowledge and understanding by answering the End of Topic Questions	4.22–4.25	259–260	
C5a.1 Orientation	Learning objectives • To revise the concept of the reactivity series • To relate the method of extraction of a metal to its reactivity • To revise the reduction of oxides by heating with carbon • To revise basic ideas of electrolysis Learning outcomes • Be able to explain how the methods of extraction of metals relate to their position in the reactivity series • Know that electrolysis is the breaking down of a compound by passing an electric current through it • Know that an electrolyte is a substance which allows the passage of an electric current when molten or dissolved in water • Understand that reduction is loss of oxygen or gain of electrons	5.1	268–269	Obtaining metals from metal oxides
C5a.2 Aluminium: extraction and uses	Learning objectives • To study the extraction of aluminium by the electrolysis of aluminium oxide • To consider the uses of aluminium in relation to its properties Learning outcomes • Be able to describe and explain the extraction of aluminium from purified aluminium oxide by electrolysis including: ○ the use of molten cryolite as a solvent and to decrease the required operating temperature ○ the need to replace the carbon anodes as they react with the oxygen being formed ○ the high cost of the electricity • Know the half-equations for the reactions of aluminium and oxide ions at the cathode and anode electrodes • Know the uses of aluminium and be able to explain these in terms of its properties	5.2–5.5	269–272	
C5a.3 Iron: extraction and uses	Learning objectives • To study the extraction of iron from iron ore in a blast furnace • To consider the uses of iron in relation to its properties	5.4–5.5	272–276	

Activity	Learning outcomes and objectives	Specification reference	Student Book pages	Practical activity
	Learning outcomes • Be able to describe and explain the main reactions involved in the extraction of iron from iron ore (haematite) using coke, limestone and air in a blast furnace • Know the uses of iron and be able to explain these in terms of its properties			
C5a.4 Consolidation and summary	Learning objectives • To review the learning points of the topic • To test understanding through answering questions Learning outcomes • Be familiar with the knowledge and understanding summarised in the End of Topic Checklist • Be able to apply this knowledge and understanding by answering the End of Topic Questions	5.1–5.5	277–279	
C5b.1 Orientation	Learning objectives • To introduce crude oil as a fossil fuel • To revise ideas on the alkane and alkene series of hydrocarbons Learning outcomes • Understand how crude oil is formed and that it is a non-renewable fossil fuel • Know that a hydrocarbon is a compound containing hydrogen and carbon only • Know that alkanes are a series of saturated hydrocarbons • Know that alkenes are a series of unsaturated hydrocarbons	5.6	280–281	
C5b.2 The fractional distillation of crude oil	Learning objectives • To study the industrial process of the fractional distillation of crude oil • To consider the nature and properties of the different substances (fractions) that can be obtained from crude oil Learning outcomes • Understand that crude oil is a mixture of hydrocarbons • Be able to describe and explain how the industrial process of fractional distillation separates crude oil into fractions • Know the names and uses of the main fractions obtained from crude oil: refinery gases, gasoline, kerosene, diesel, fuel oil, bitumen • Be able to describe the trend in boiling point and viscosity of the main fractions	5.7–5.9	281–283	Demonstration: the fractional distillation of crude oil
C5b.3 Cracking crude oil fractions	Learning objectives • To introduce the process of cracking hydrocarbon chains • To explain the reasons for using the cracking process	5.12–5.13	283–285	Cracking liquid paraffin

Activity	Learning outcomes and objectives	Specification reference	Student Book pages	Practical activity
	Learning outcomes • Understand that the fractional distillation of crude oil produces more long-chain hydrocarbons than can be used and fewer short-chain hydrocarbons than are needed and so cracking is necessary • Be able to describe how long-chain hydrocarbons can be converted into shorter chain hydrocarbons by catalytic cracking • Know that oil fractions are cracked using catalysts of silica or alumina at a temperature of between 600 and 700 °C			
C5b.4 The combustion of fuels and acid rain	Learning objectives • To consider some of the problems of pollution caused by burning hydrocarbon fuels • To consider specifically the problems of acid rain Learning outcomes • Understand that the incomplete combustion of hydrocarbon fuels may produce carbon monoxide • Be able to explain that carbon monoxide is poisonous because it reduces the capacity of the blood to carry oxygen • Understand that in car engines the temperature is hot enough for nitrogen and oxygen to combine and form nitrogen oxides • Understand that nitrogen oxides and sulfur dioxide are pollutant gases which contribute to acid rain, and be able to describe the problems caused by acid rain	5.10–5.11	285–290	
C5b.5 Consolidation and summary	Learning objectives • To review the learning points of the topic • To test understanding through answering questions • To consider sources of energy for the future Learning outcomes • Be able to explain the limitations of using fossil fuels as sources of energy • Be familiar with the knowledge and understanding summarised in the End of Topic Checklist • Be able to apply this knowledge and understanding by answering End of Topic Questions	5.6–5.13	289–294	
C5c.1 Orientation	Learning objectives • To revise the structures of the alkanes and alkenes • To revise the addition reactions of the alkenes • To introduce the terms monomer, polymer and polymerisation Learning outcomes • Understand that alkenes are much more reactive than alkanes • Understand that alkenes undergo addition reactions	5.14	295–296	

Activity	Learning outcomes and objectives	Specification reference	Student Book pages	Practical activity
	• Know that a monomer is a molecule that can combine with other molecules to form a polymer			
C5c.2 Addition polymerisation	Learning objectives • To introduce and explain the nature of addition polymerisation. • To consider the uses of some of the common examples of addition polymers. • To consider the problems of recycling and disposing of used plastic material. Learning outcomes • Understand that an addition polymer is formed by joining up many small molecules called monomers. • Be able to draw repeat topics of addition polymers from the structural formula of the monomer, such as poly(ethene), poly(propene) and poly(chloroethene). • Be able to deduce the structural formula of a monomer from the repeat topic of an addition polymer. • Be able to describe some of the uses for polymers, including poly(ethene), poly(propene) and poly(chloroethene). • Be able to explain that addition polymers are hard to dispose of as their inertness means that they do not easily biodegrade.	5.14–5.16	295–297	Identifying addition polymers
C5c.3 Condensation polymerisation	Learning objectives • To study the formation of condensation polymers • To consider the differences between condensation and addition polymers Learning outcomes • Understand that some polymers, such as nylon, form by a different process called condensation polymerisation • Understand that when condensation polymers are formed small molecules, such as water, are released	5.17–5.20	297–299	Demonstration: condensation polymerisation
C5c.4 Consolidation and summary	Learning objectives • To review the learning points of the topic • To test understanding through answering questions Learning outcomes • Be familiar with the knowledge and understanding summarised in the End of Topic Checklist • Be able to apply this knowledge and understanding by answering the End of Topic Questions	5.14–5.20	300–301	
C5d.1 Orientation	Learning objectives • To revise ideas on reaction rate and equilibria • To revise understanding of electrolysis • To revise redox reactions Learning outcomes • Know the factors that increase the rate of a reaction	5.21	302	

Activity	Learning outcomes and objectives	Specification reference	Student Book pages	Practical activity
	• Know the factors that affect the equilibrium in a reversible reaction • Understand the meaning of the term electrolysis • Understand oxidation and reduction in terms of loss or gain of electrons or loss or gain of oxygen			
C5d.2 The manufacture of ammonia	Learning objectives • To study the Haber process for the manufacture of ammonia • To study the use of ammonia in the manufacture of nitric acid • To consider some of the important uses of ammonia Learning outcomes • Be able to describe how ammonia is manufactured from nitrogen and hydrogen in the Haber process • Understand that the nitrogen is obtained from the air and the hydrogen from natural gas or the cracking of hydrocarbons • Be able to describe the conditions used in the Haber process: a temperature of 450 °C, a pressure of 200 atmospheres, an iron catalyst the ammonia is liquefied and the nitrogen and hydrogen are recycled • Understand that the choice of conditions in the Haber process provides a balance between maximising the yield of ammonia and the rate of the reaction • Be able to describe that ammonia is used in the manufacture of nitric acid and fertilisers	5.21–5.24	302–304	
C5d.3 The manufacture of sulfuric acid	Learning objectives • To study the reactions of the contact process used in the manufacture of sulfuric acid • To consider some of the important uses of sulfuric acid Learning outcomes • Be able to describe how sulfuric acid is made in the contact process, starting from sulfur and air, but the main reaction involves the oxidation of sulfur dioxide to sulfur trioxide • Know that the essential conditions in the contact process include: a temperature of 450 °C a pressure of about 2 atmospheres a catalyst of vanadium(V) oxide • Be able to describe the use of sulfuric acid in the manufacture of detergents, fertilisers and paints	5.25–5.27	304–306	
C5d.4 The manufacture of sodium hydroxide and chlorine	Learning objectives • To study the electrolysis of concentrated sodium chloride in a diaphragm cell • To use half-equations to summarise the reactions taking place at the electrodes • To consider some of the important uses of sodium hydroxide and chlorine Learning outcomes • Be able to describe how sodium hydroxide and chlorine are manufactured by the electrolysis of concentrated	5.28–5.30	307–310	The electrolysis of sodium chloride solution

Activity	Learning outcomes and objectives	Specification reference	Student Book pages	Practical activity
	sodium chloride solution in a diaphragm cell • Be able to write half-equations for the reactions taking place at the electrodes in the diaphragm cell • Be able to describe the important uses of sodium hydroxide, including the manufacture of bleach, paper and soap • Be able to describe the important uses of chlorine, including sterilising water supplies, and in the manufacture of bleach and hydrochloric acid			
C5d.5 Consolidation and summary	Learning objectives • To review the learning points of the topic • To test understanding through answering questions Learning outcomes • Be familiar with the knowledge and understanding summarised in the End of Topic Checklist • Be able to apply this knowledge and understanding by answering the End of Topic Questions	5.21–5.30	309–313	